Lecture Notes in Control and Information Sciences

Edited by M. Thoma

Vol. 43: Stochastic Differential Systems
Proceedings of the 2nd Bad Honnef Conference
of the SFB 72 of the DFG at the University of Bonn
June 28 – July 2, 1982
Edited by M. Kohlmann and N. Christopeit
XII, 377 pages. 1982.

Vol. 44: Analysis and Optimization of Systems
Proceedings of the Fifth International
Conference on Analysis and Optimization of Systems
Versailles. December 14–17, 1982
Edited by A. Bensoussan and J. L. Lions
XV, 987 pages, 1982

Vol. 45: M. Arató
Linear Stochastic Systems
with Constant Coefficients
A Statistical Approach
IX, 309 pages. 1982

Vol. 46: Time-Scale Modeling of Dynamic Networks
with Applications to Power Systems
Edited by J. H. Chow
X, 218 pages. 1982

Vol. 47: P. A. Ioannou, P. V. Kokotovic
Adaptive Systems with Reduced Models
V, 162 pages. 1983

Vol. 48: Yaakov Yavin
Feedback Strategies for Partially
Observable Stochastic Systems
VI, 233 pages, 1983

Vol. 49: Theory and Application of Random Fields
Proceedings of the IFIP-WG 7/1
Working Conference
held under the joint auspices of the
Indian Statistical Institute
Bangalore, India, January 1982
Edited by G. Kallianpur
VI. 290 pages. 1983

Vol. 50: M. Papageorgiou
Applications of Automatic Control Concepts
to Traffic Flow Modeling and Control
IX, 186 pages. 1983

Vol. 51: Z. Nahorski, H.F. Ravn, R.V.V. Vidal
Optimization of Discrete Time Systems
The Upper Boundary Approach
V, 137 pages 1983

Vol. 52: A. L. Dontchev
Perturbations, Approximations and Sensitivity Analysis
of Optimal Control Systems
IV, 158 pages. 1983

Vol. 53: Liu Chen Hui
General Decoupling Theory of Multivariable
Process Control Systems
XI, 474 pages. 1983

Vol. 54: Control Theory for Distributed
Parameter Systems and Applications
Edited by F. Kappel, K. Kunisch,
W. Schappacher
VII, 245 pages. 1983.

Vol. 55: Ganti Prasada Rao
Piecewise Constant Orthogonal Functions
and Their Application to Systems and Control
VII, 254 pages. 1983.

Vol. 56: Dines Chandra Saha, Ganti Prasada Rao
Identification of Continuous
Dynamical Systems
The Poisson Moment Functional
(PMF) Approach
IX, 158 pages. 1983.

Vol. 57: T. Söderström, P. G. Stoica
Instrumental Variable Methods
for System Identification
VII, 243 pages. 1983.

Vol. 58: Mathematical Theory of
Networks and Systems
Proceedings of the MTNS-83 International
Symposium
Beer Sheva, Israel, June 20–24, 1983
Edited by P. A. Fuhrmann
X, 906 pages. 1984

Vol. 59: System Modelling and Optimization
Proceedings of the 11th IFIP Conference
Copenhagen, Denmark, July 25-29, 1983
Edited by P. Thoft-Christensen
IX, 892 pages. 1984

Vol. 60: Modelling and Performance
Evaluation Methodology
Proceedings of the International Seminar
Paris, France, January 24–26, 1983
Edited by F. Bacelli and G. Fayolle
VII, 655 pages. 1984

Vol. 61: Filtering and Control of Random
Processes
Proceedings of the E.N.S.T.-C.N.E.T. Colloquium
Paris, France, February 23–24, 1983
Edited by H. Korezlioglu, G. Mazziotto, and
J. Szpirglas
V, 325 pages. 1984

For information about Vols. 1–42 please contact your bookseller or Springer-Verlag.

Lecture Notes in Control and Information Sciences

Edited by M. Thoma and A. Wyner

104

J. S. Freudenberg, D. P. Looze

Frequency Domain Properties of Scalar and Multivariable Feedback Systems

Springer-Verlag
Berlin Heidelberg GmbH

Authors
James Scott Freudenberg
4213 EECS Bldg., EECS Dept.
University of Michigan
Ann Arbor, MI 48109-2122
USA

Douglas Paul Looze
Dept. of Electrical and Computer Eng.
University of Massachusetts
Amherst, MA 01002
USA

ISBN 978-3-540-18869-8

Library of Congress Cataloging in Publication Data
Freudenberg, J. S. (James Scott),
Frequency domain properties of scalar and multivariable feedback systems /
J. S. Freudenberg, D. P. Looze.
(Lecture notes in control and information sciences 104)
Bibliography: p.
ISBN 978-3-540-18869-8 ISBN 978-3-540-38839-5 (eBook)
DOI 10.1007/978-3-540-38839-5

1. Feedback control systems.
I. Looze, D. P. (Douglas Paul), II. Title, III. Series.
TJ216.F74 1988
629.8'3--dc19 88-1997

Originally published by Springer-Verlag Berlin Heidelberg New York in 1988

2161/3020-543210

To my grandparents,
J.S. Freudenberg

To my wife,
D.P. Looze

Preface

The research reported in this monograph was initiated when both authors were at the University of Illinois; and continued when the second author moved to ALPHATECH and the University of Massachusetts and the first author moved to the University of Michigan. Although our research program is still in development, we feel that enough has been accomplished to warrant a comprehensive presentation of our approach.

We would like to acknowledge helpful discussions with many colleagues. In particular, the first author is grateful for the opportunity to interact with the personnel of the Honeywell Systems and Research Center during the summer of 1982. Discussions with Bruce Francis and George Zames have also proven quite useful. An enormous debt of gratitude is owed to Professor Stephanie Alexander, of the Mathematics Department at the University of Illinois. Without her enthusiastic assistance, the work reported in Chapters 8-11 would never have been completed. The word-processing is due to Elizabeth Olsen and Virginia Folsom of the University of Michigan. Finally, the work of the first author was supported in part by the United States National Science Foundation, under Grant ECS 8504558.

TABLE OF CONTENTS

CHAPTER 1

INTRODUCTION

The process of feedback design is necessarily performed in the presence of hard constraints upon properties of the closed-loop system. Consider, for example, the structure imposed by the existence of the feedback loop itself. On the one hand, incorporating knowledge of the system output into the control law can allow the effects of disturbances and small modelling errors to be reduced. On the other hand, this process requires the use of an inevitably noisy measurement, and may also lead to instability in the presence of modelling errors. Hence the very nature of a feedback loop constrains the properties of the resulting design, and this is true whether the system is linear or nonlinear, time-varying or time-invariant.

When a system can be approximately described by a linear time-invariant model, then many of the constraints upon feedback properties can be expressed mathematically. To summarize the reasons for this, recall that the input-output behavior of a system which is linear, time-invariant, stable, and causal is determined completely by its steady-state response to sinusoidal inputs (e.g. [ZaD63], p. 418). Hence, the design of a feedback law for a linear time-invariant system is equivalent to manipulating its frequency response properties. For scalar systems (i.e., those with a single input and output) this amounts to shaping a complex-valued scalar function of the frequency variable $s=j\omega$. However, as Bode emphasized [Bod45], a complex-valued function has a realization as the frequency response of a physical system only if rather stringent constraints are satisfied. These *realizability* constraints essentially show that the frequency response of a stable, causal, linear, time-invariant system cannot be manipulated independently in different frequency ranges. The resulting design constraints are described mathematically using the classic Bode integrals. Constraints upon feedback design may also be imposed by properties specific to the system being controlled. For example, plants which are nonminimum phase, unstable, or have bandwidth limitations are all more difficult to control than if these properties were not present.

The constraints imposed upon the feedback properties of a linear time-invariant system manifest themselves as tradeoffs that must be performed between mutually exclusive design goals. The importance of tradeoffs in feedback design has long been recognized. Indeed, for scalar systems, a useful theory describing tradeoffs has been developed and incorporated into classical design methodology ([Bod45, Hor63]). A number of important results concerned with design constraints in multivariable systems (i.e., those with multiple inputs and/or outputs) have also been developed; as the reference cited in Chapter 5 will attest, this area has received considerable attention in recent years. Nevertheless, the theory of design constraints in multivariable feedback systems remains relatively incomplete, and this limits the efficiency of current multivariable design methodologies.

Our goal in this monograph is to continue the study of constraints imposed upon feedback system design. As we restrict our attention to systems that are linear and time-invariant, we are able to conduct our study solely in the frequency domain. This is consistent with our overall strategy of identifying concepts important in the classical theory of scalar systems and extending these to a multivariable setting. Hence we first need to clarify how the existence of constraints affects the design of a feedback system, and to determine the tools needed to incorporate knowledge of these constraints into the design process. Although we will not develop a design methodology *per se*, the results of our study of feedback system properties should prove useful in *any* methodology.

Constraints upon feedback properties affect all stages of system design. For example, when it is known that feedback will eventually be needed to control a system, then knowledge of the limitations of feedback may be incorporated directly into the design of the plant, so that the difficulty of the eventual control problem is not excessive. Once the plant is given, it is necessary to develop a set of goals, or specifications, which the feedback design should satisfy. Typically, these specifications are stated as requirements upon the properties of *closed loop transfer functions* governing the response of signals within the feedback loop to exogeneous inputs. It is thus important to know how constraints upon feedback properties affect these closed loop transfer functions, in order that design specifications may be developed which are compatible with the limitations inherent in the system.

Once a design specification has been constructed, it is then necessary to apply a synthesis technique to develop a candidate design. This is generally an iterative procedure, and several candidate compensators must be examined before a final choice is made. At each iteration, the properties of the resulting design must be analyzed; if the specification is not satisfied, then the candidate design must be modified so that the specification is more nearly satisfied. This process is expedited if the relations between the *design parameters* available in the synthesis technique and the properties of the resulting design are well understood. Clearly, one's ability to manipulate feedback system properties using the available design parameters will be limited by all the constraints discussed above. If the original specification is not achievable, then knowledge of the inherent feedback design tradeoffs is useful in obtaining a design which violates the specification no more than necessary. On the other hand, it may be possible to tighten portions of the specification; the same body of knowledge is useful in this task.

It is clear that a well developed understanding of feedback system properties can only serve to increase the efficiency of the design process. To be most useful, this theory must be expressed in two forms. First, an explicit and precise mathematical theory of feedback properties must be obtained; in particular, it is necessary to understand how constraints due to feedback structure, to realizability conditions, and to plant properties limit the behavior of the closed loop system. However, from a practical standpoint, the purpose of having a theory is to save time in the design process, and thus merely having a set of theorems and proofs is not enough. Also needed are *heuristics*, or rules of thumb, to aid in the design. Although heuristics are most appealing when they are based upon a rigorous underlying theory, they themselves may be less precise. Indeed, heuristics typically sacrifice rigor and precision for speed and insight. By describing, in qualitative terms, how feedback systems must behave "most" of the time, heuristics can provide information useful to the designer without having to resort to the complexity of the complete underlying theory. Hence, heuristics can allow a rapid assessment of the tradeoffs and limitations inherent in a given feedback design problem, can aid in the development of design specification compatible with these limitations, and can enable a designer to quickly prototype a compensator which is in the "ball park" of being satisfactory.

It is undeniable that classical methodology is successful in addressing an important, albeit limited, class of feedback design problems. This observation motivates us to reflect upon the reasons why classical methodology is successful, with the viewpoint that these features should be incorporated, if at all possible, in a multivariable design methodology. Classical design is greatly aided by the existence of a rigorous mathematical theory of feedback systems, especially pertaining to the limitations imposed upon feedback properties [Nyq32, Bod45]. Furthermore, there exists an extensive set of heuristics[Bod45, Hor63], which are deeply embedded in the terminology and the graphical analysis and design methods of classical control. Although it is likely that a successful multivariable design methodology will also incorporate both a rigorous theory of feedback behavior as well as heuristics to speed the design process, the specific form that these features will take is at present undetermined.

Other useful features of classical design, such as those inherently related to the classical graphical analysis methods, may be limited to scalar systems. Using graphical methods, it is relatively easy to translate design specification from the closed loop transfer functions, where they are most naturally posed, to the open loop transfer function. Since classical methodology uses open loop gain and phase as design parameters, restating the specifications in this way greatly facilitates compensator design. In addition, there are a number of design limitations, such as the additional phase lag contributed by a nonminimum phase plant zero, that appear more amenable (at least superficially) to analysis using the open loop transfer function. Unfortunately, the relation between open and closed loop properties of multivariable feedback systems is not so transparent, and the role of phase in a multivariable context is not well understood. Together, these facts render problematic the generalization of many classical concepts to multivariable systems. Indeed, only in special cases, such as when a multivariable system consists of a set of nearly decoupled scalar systems, can the classical analysis and design methodologies be reliably applied. When coupling cannot be neglected, it is possible for a multivariable system to exhibit properties that have no scalar analogue, and that cannot be analyzed using only scalar methods. In particular, the properties of a multivariable system can exhibit a strong spatial, or *directional,* dependence in addition to the frequency dependence shared with scalar systems. Far too little is currently known concerning the mechanisms by which directionality can affect feedback properties. In particular, little is known about the effect that directionality has upon the design limitations imposed by feedback structure, realizability constraints, and plant properties.

The number of degrees of freedom present in a multivariable design problem undoubtedly precludes development of any design technique as transparent as the classical methodology for scalar systems. One popular approach to multivariable design is to reformulate the design problem as one of *optimizing* a cost function representing the design specifications imposed upon the closed loop transfer functions. The design parameters in such problems are weighting functions used to reflect the relative importance of different design goals (see, for example, [DoS81, SLH81, StA84, FrD85, Hel85, Kwa85, Doy85, Boy86], to name but a few). If it were possible to develop an optimization methodology whose objective function *accurately* incorporated *all* relevant design goals, including an accurate ordering of their relative importance, then the need for an explicit theory of feedback design constraints might be reduced. In the ideal methodology, one would merely have to formulate the design problem carefully, solve the resulting optimization problem, and then implement the optimal design. At present, this goal has not been attained, and as a result, the optimization-based methodologies have achieved their greatest successes when used as tools for iterative design. When an optimization methodology is used in this context, the weightings in the cost function are typically adjusted until either a satisfactory design is achieved or it is concluded that such a design does not exist. Hence the iterative nature of classical design methodology is retained, but the design parameters are the weighting functions, instead of open loop gain and phase. Since the weightings are applied directly to the closed loop transfer functions upon which design specifications are imposed, this would appear to be quite an advantage. However, the limitations and tradeoffs imposed by the various design constraints are still present, albeit only implicitly stated. It follows that the weighting function must be selected to reflect, not only the *desired* properties of the feedback system, but also the fact that these desired properties must be compatible with the inherent design limitations. As we have seen, the theory of feedback design limitations in multivariable systems is incomplete, and this fact thus complicates the application of current multivariable design methodologies.

Our primary goal in this monograph is to study the frequency-response properties of linear time-invariant multivariable feedback systems. We are especially interested in limitations imposed upon these properties by the constraints of a feedback structure, of realizability, and of properties of the plant to be controlled. As a prologue, we shall first review some well-known properties of scalar feedback systems and discuss some recently obtained results. Our eventual goal is to develop a rigorous mathematical theory describing properties of multivariable feedback systems and to derive a set of heuristics, based upon that theory, to aid in design. Although the theoretical development remains incomplete, we do obtain significant insights into the properties of multivariable feedback systems. Similarly, although a thorough understanding of design implications is not yet available, a number of heuristic interpretations are provided which should prove useful in design.

We are interested in three physical sources of limitations that impose upon properties of a feedback design. Each of these limitations is manifested as a design tradeoff, and it is convenient to group these tradeoffs into two categories. *Algebraic* design tradeoffs, so-called because the associated mathematical equations are algebraic, must be performed between different feedback properties at the same frequency. Such tradeoffs typically arise due to the structure of the feedback loop; an example is the well-known tradeoff which exists at each frequency between response to disturbances and response to noise. *Analytic* design tradeoffs, on the other hand, must be performed between system properties in *different* frequency ranges. They are a direct consequence of the constraints that realizability imposes upon a linear time-invariant system. These tradeoffs may be expressed mathematically using the fact that the transfer function of a stable, causal, linear time-invariant system is analytic in the closed right half plane (hence the adjective "analytic"). Just as physical realizability imposes strict limitations upon the properties of a linear time-invariant system, analyticity imposes considerable mathematical structure upon a transfer function. Hence, the tools of complex variable theory may be used to study transfer functions, and the results may then be interpreted in the context of linear systems. Finally, we remark that the sets of algebraic and analytic constraints are not mutually exclusive, and there exist design constraints which may be stated either way. In particular, there are a number of constraints due to structural features of the plant, such as unstable poles and nonminimum phase zeros, which may be stated either algebraically, at points in the open right half of the complex frequency plane, or analytically, as integrals along the $j\omega$-axis, where design specifications are imposed. The latter interpretation is useful, since it lends insight into the design tradeoffs that must be performed between system properties in different frequency ranges.

Our approach to multivariable feedback systems will attempt to parallel, to the extent possible, the classical approach to analysis of scalar systems. However, there do exist a number of difficulties associated with this program, and one of these deserves immediate comment. Earlier, we remarked that classical methodologies focussed upon open loop gain and phase, both for the analysis of feedback properties, as well as for use as design parameters. That this procedure can be successful is due to the ease with which feedback properties of a scalar system may be related to open loop gain and phase. As we mentioned, these relations are quite problematic in a multivariable setting. On the one hand, since the closed loop transfer functions retain their useful interpretations for multivariable systems, it appears plausible that multivariable design problems should be posed in terms of these transfer functions. Indeed, many, if not most, multivariable design methodologies take this approach. Yet there remains a substantial body of heuristics stated in terms of open loop gain and phase. For example, there is a well-known design tradeoff, described qualitatively by the Bode gain-phase relation, which limits the rate at which open loop gain may be rolled off near crossover frequency. Another example is the difficulty due to the additional phase lag contributed by a nonminimum phase plant zero. Hence, it would seem prudent to translate that portion of classical theory and heuristics which is stated in terms of the open loop transfer function into equivalent statements in terms of the closed loop transfer functions. Once this task is completed for scalar systems, the results could presumably be generalized to the multivariable case.

Alternately, one could note that the failure of classical design techniques to extend to multivariable systems does not necessarily preclude the extension of classical analysis tools and heuristics. Hence it may prove useful to study the relation between open and closed loop properties of multivariable systems, if only to shed light on design limitations imposed by properties such as open loop bandwidth constraints and nonminimum phase plant zeros. It is our opinion that both these lines of inquiry should be pursued. Indeed, the tools we develop for the multivariable generalization of the gain-phase relation, an open loop result, should prove equally useful in studying generalizations of other integral relations which constrain closed loop properties directly.

The monograph is organized into four sections. In Chapters 2 through 4, we examine constraints imposed upon the properties of scalar feedback systems, and discuss briefly how classical methodology dealt with the associated design tradeoffs. Chapter 2 provides background material by motivating and formulating the scalar feedback design problem. We review portions of classical feedback theory, note the distinction between algebraic and analytic design tradeoffs, and discuss several heuristics useful in classical analysis and design. We note how design specifications imposed upon the closed loop transfer functions may be translated into specifications upon open loop gain and phase. Chapter 3 continues this discussion by examining limitations upon system properties which, although long recognized, have only recently been stated in the form of analytic design tradeoffs. Chapter 4 concludes the discussion of scalar systems by examining how constraints upon feedback properties were incorporated into classical analysis and design techniques.

Our development of multivariable extensions is divided into two portions, the first of which comprises Chapters 5 through 7. In Chapter 5, we review prior work on multivariable extensions of classical feedback concepts. In particular, we describe heuristics appropriate for multivariable systems that behave like scalar systems in the sense that directionality is not an issue. Loosely speaking, this will be the case whenever all loops of the system possess the same qualitative properties at each frequency. For such systems, the generalization of classical concepts is relatively straightforward. However, there exist many multivariable systems that cannot be analyzed using simple extensions of scalar concepts. In certain cases, the scalar ideas are still capable of extension, provided that directionality considerations are taken properly into account. For example, the singular value decomposition can be used to extend the notion of gain, including the use of the singular vectors to describe a set of directions corresponding to each singular value "gain". In Chapter 6, we discuss some of the difficulties associated with the extension of phase to multivariable systems. By analyzing multivariable signals, rather than systems, we obtain an appealing measure of multivariable phase *difference*. This notion of phase takes directionality issues into account and also retains some of the useful properties of scalar phase. In Chapter 7, we use the multivariable extensions of gain and phase, together with directional information, to extend some classical heuristics to a multivariable setting. Specifically, we study an algebraic design tradeoff by analyzing the relation between open and closed loop properties of a multivariable system at a fixed frequency.

In Chapters 8 through 11 we develop mathematical tools for studying analytic design tradeoffs in multivariable systems. This section of the monograph culminates in Chapter 11, where we develop a multivariable analogue of the classical Bode gain-phase relation. The role of gain in this generalization is played by a singular value, and that of phase by a measure of phase difference between the associated pair of singular vectors. Directionality properties of the system must also be considered; in particular, the fact that the singular subspaces may vary throughout the right half plane is seen to impact design tradeoffs. The considerable background needed to reach this result is the subject of Chapters 8 through 10. These intermediate results are of independent interest, as they should provide a framework for extending the other scalar integral relations to multivariable systems. The tool used to study analytic design tradeoffs in the scalar case is complex variable theory. That tool is also useful in multivariable systems, but cannot be applied directly. In Chapter 8 we discuss the reasons why this is so; essentially, the fact that directionality properties change with frequency implies that the singular values and other coordinates of interest do not constitute complex analytic functions. To study this phenomenon further, it is necessary to describe the geometric structure of the singular vectors. This we do in Chapter 9, and obtain a precise method of distinguishing between the "phase" of a vector signal and the direction in which it lies. This framework is then used, in Chapter 10, to derive a set of differential equations relating the singular values and vectors. These equations, reminiscent of the familiar Cauchy-Riemann equations, provide the basis for deriving the multivariable integral relation in Chapter 11.

The final results of the monograph are contained in Chapter 12. There, we use the results from Chapters 7 and 11 to analyze properties of a multivariable system exhibiting distinctly nonscalar behavior. Specifically, the system is required to have a large rate of gain decrease in one of its loops; the nonscalar behavior arises since the associated sensitivity function has two peaks, each of which is substantially smaller than the single peak which would be present in a scalar system with this rate of gain decrease. Section 13 concludes the monograph by describing directions for further research.

CHAPTER 2

REVIEW OF CLASSICAL RESULTS

2.1. Introduction

The purpose of this chapter is to review some important ideas from the classical theory of linear time-invariant scalar feedback systems. We shall focus on those ideas from the scalar theory which we wish to extend to multivariable systems in later chapters of this monograph. Specifically, we will concentrate on formulating feedback design goals as specifications imposed upon closed loop transfer functions, and then map these into equivalent specifications upon the open loop transfer function.

It is well-known that feedback may be used to stabilize an unstable system, to reduce the response of the system to disturbances, and to reduce the effect of plant parameter variations and modelling errors upon the system output. In addition to these potential benefits of using feedback there are also potential disadvantages. In particular, one must ensure that response to sensor noise is not excessively large and that the system is robustly stable. All these properties of a feedback system may be quantified and evaluated using certain closed loop transfer functions.

For a variety of reasons, not all performance specifications are achievable. Among these are limitations inherent in the structure of a feedback loop and limitations due to properties of linear time-invariant systems. Still other design limitations may be imposed by properties of the plant to be controlled. Frequently, these limitations appear as *tradeoffs* which must be performed in the design process. It will be convenient in this monograph to divide these tradeoffs into two categories. An *algebraic* tradeoff is quantified by an algebraic equation obtained by evaluating a transfer function identity at a complex frequency of interest. Such a tradeoff must be performed between different properties of a feedback system at the *same* frequency. *Analytic* tradeoffs are quantified by integral equations derived using complex variable theory. These tradeoffs are characterized by the fact that they must be performed between system properties in *different* frequency ranges.

Classical design methodologies proceeded by "shaping" the gain and phase of the open loop transfer function to modify the feedback properties of the closed loop system. This strategy is successful only because the relations between open and closed loop system properties are relatively well understood. Hence, specifications imposed upon closed loop transfer functions may be mapped into equivalent specifications upon the open loop transfer function. In addition, certain of the analytic design tradeoffs may be expressed in terms of open loop gain and phase. The relations between the open and closed loop system properties can be expressed as a set of heuristics, or "rules of thumb," which greatly facilitate the design process.

The organization of this chapter reflects the important role that the relationship between open and closed loop transfer functions played in classical feedback design. Section 2.2 presents the classical design problem for a scalar system, and examines the properties of the closed loop transfer functions that are important to satisfactory system performance. In Section 2.3, we discuss how design specifications can be translated into frequency dependent bounds upon the magnitudes of the closed loop transfer functions. Two tradeoffs imposed by algebraic design constraints are also examined. Section 2.4 is devoted to a review of heuristics useful in approximating feedback properties of the closed loop system from the properties of the open loop transfer function. In Section 2.5, Bode's gain-phase relation is analyzed to display the analytic constraint it imposes upon open loop gain and phase and in turn open feedback properties. Finally, in Section 2.6 the classical results are summarized by identifying three general areas into which they may be divided. This classification provides a convenient framework for discussing the multivariable extensions pursued in later chapters.

2.2: Quantification of Scalar Feedback Performance

Consider the linear time-invariant feedback system shown in Figure 2.2.1. The transfer functions $P(s)$ and $F(s)$ are those of the plant and compensator, respectively. The signal $r(s)$ is a reference or command input, $y(s)$ is the system output, $d(s)$ is a disturbance input, and $n(s)$ is sensor noise. Three additional transfer functions are essential in studying properties of this system. Define the *open loop transfer function*

$$L(s) \triangleq P(s)F(s) \tag{2.2.1}$$

the *sensitivity function*

$$S(s) \triangleq [1+L(s)]^{-1} \tag{2.2.2}$$

and the *complementary sensitivity function* ([Kwa83, Kwa85])

$$T(s) \triangleq L(s)[1+L(s)]^{-1} \tag{2.2.3}$$

The output of the system can be decomposed into the sum of three terms:

$$y(s) = y_r(s) + y_d(s) + y_n(s) \ , \tag{2.2.4}$$

where

$$y_r(s) = S(s)P(s)r(s) \tag{2.2.5}$$

$$y_d(s) = S(s)d(s) \tag{2.2.6}$$

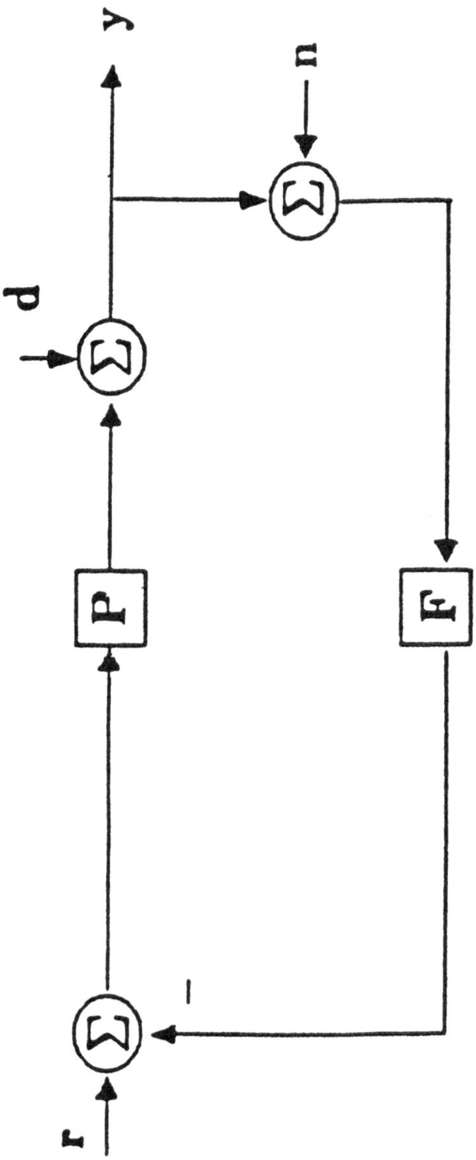

Figure 2.2.1: Linear Time Invariant Feedback System

and

$$y_n(s) = -T(s)n(s) \tag{2.2.7}$$

denote, respectively, the responses of the system to commands, disturbances, and noise. Similarly, the input to the plant can be decomposed as

$$u(s) = u_r(s) + u_d(s) + u_n(s) \quad , \tag{2.2.8}$$

where

$$u_r(s) = S(s)\, r(s) \tag{2.2.9}$$

$$u_d(s) = -S(s)\, F(s)\, d(s) \tag{2.2.10}$$

and

$$u_n(s) = -S(s)\, F(s)\, n(s) \quad . \tag{2.2.11}$$

Assume that the open loop system with transfer function (2.2.1) is free of unstable hidden modes. Then the feedback system will be *stable* if the sensitivity function is bounded in the closed right half plane. Note that the assumption of no unstable hidden modes implies that closed right half plane poles and zeros of $P(s)$ and $F(s)$ must appear with at least the same multiplicity in $L(s)$.

If the feedback system is stable, then the steady-state response of the system output to disturbances $e^{j\omega t}$ is governed by the sensitivity function: $y_d(t) \rightarrow S(j\omega)e^{j\omega t}$. Hence the response to disturbances with frequency ω can be made small by requiring that $|S(j\omega)| \ll 1$. Indeed, since the disturbance response of a system without feedback is unity at all frequencies, it follows that feedback improves the disturbance response at any frequency for which $|S(j\omega)| < 1$. Use of feedback, however, implies that the effects of noise upon the system output must be considered. Steady state response to noise signals $e^{j\omega t}$ is governed by the complementary sensitivity function: $y_n(t) = -T(j\omega)e^{j\omega t}$. Noise response can be made small, therefore, by requiring that $|T(j\omega)| \ll 1$.

Feedback can also be beneficial in reducing the effects of plant modelling errors and parameter variations. The effect of such uncertainty upon the system output may be assessed from [CrP64]

$$E_c(s) = S'(s)E_o(s) \quad , \tag{2.2.12}$$

$$S'(s) \triangleq [1+P'(s)F(s)]^{-1} \quad , \tag{2.2.13}$$

where $P'(s)$ is the transfer function of the *true* plant, rather than the nominal model, and $E_c(s)$ and $E_o(s)$ are the deviations in the outputs of nominally equivalent closed and open loop designs caused by the modelling error $P'(s) - P(s)$. Equation (2.2.12) allows the effect of feedback upon modelling error to be analyzed on a frequency by frequency basis. At each frequency for which $|S'(j\omega)| < 1$, the feedback system possesses the sensitivity reduction property; i.e., the error in the output of the system due to plant model errors is less than if no feedback were present. This condition is difficult to use directly as a design criterion, since its evaluation depends upon the true (unknown) plant rather than the model. However, requiring $|S(j\omega)| < 1$ does imply that $|S'(j\omega)| < 1$ for "sufficiently small" model errors (c.f. [Fra78]). Recent results [DWS82] are useful in testing whether the condition $|S'(j\omega)| < 1$ is satisfied for large model errors.

In addition to the properties discussed above, the sensitivity and complementary sensitivity functions each characterize stability robustness of the system against a particular class of plant uncertainty (e.g., see[DWS82], Table 1.). One particularly important class of uncertainty arises in practice because models of linear systems inevitably deteriorate in accuracy at high frequencies[DoS81]. Typically, unmodelled high frequency dynamics can cause large relative uncertainty in plant gain and complete uncertainty in phase. This type of uncertainty can be described by modelling the difference between the true plant and the model by *multiplicative* uncertainty:

$$P'(s) = (1+\Delta(s))P(s) \quad . \tag{2.2.14}$$

The perturbation $\Delta(s)$ is assumed to lie in a set of stable transfer functions satisfying a frequency-dependent magnitude bound:

$$\begin{aligned} D(s) \triangleq \{\Delta(s) : \ & \Delta(s) \text{ is stable and} \\ & |\Delta(j\omega)| \le M(\omega) \, , \ \forall \ \omega\} \end{aligned} \tag{2.2.15}$$

If the only information available about the uncertainty is (2.2.14-15) then $\Delta(s)$ is termed *unstructured* uncertainty. In that case, a necessary and sufficient condition for the feedback system to be stable for all plants described by (2.2.14-15) is that the system be stable when $\Delta = 0$ and that the complementary sensitivity function satisfy the bound [DoS81]

$$|T(j\omega)| < 1/M(\omega) \quad , \quad \forall \ \omega \quad . \tag{2.2.16}$$

Typically, system uncertainty is reasonably well modelled at high frequencies using a bound $M(\omega)$ which becomes increasingly large. Thus the presence of high frequency uncertainty forces the complementary sensitivity function to become small to insure that the system is robustly stable.

Note that the assumed uncertainty description contains no information about the phase of the uncertainty. If such information is available, the uncertainty is termed *structured* and the bound (2.2.16) is only a *sufficient* condition for stability robustness. For a variety of reasons, characterizing system uncertainty as unstructured and multiplicative may not be appropriate, especially at low frequencies. At high frequencies, however, the assumption is useful for many scalar design problems.

2.3: Closed Loop Transfer Functions and Design Specifications

In the previous section we saw that the sensitivity and complementary sensitivity functions each characterize important properties of a feedback system. This fact motivates stating design specifications directly as frequency-dependent bounds upon the magnitudes of these functions. Hence, we typically require that

$$|S(j\omega)| \leq M_S(\omega) \quad , \quad \forall \ \omega \tag{2.3.1}$$

and

$$|T(j\omega)| \leq M_T(\omega) \quad , \quad \forall \ \omega \quad . \tag{2.3.2}$$

The bounds $M_S(\omega)$ and $M_T(\omega)$ will generally depend upon the size of the disturbance and noise signals, the level of plant uncertainty, and the extent to which the effect of these phenomena upon the system output must be diminished.

Feedback systems must be designed to satisfy many different objectives, and frequently these objectives are mutually incompatible. It is therefore important to understand when a given design specification is achievable and when it must be relaxed by making tradeoffs between conflicting design goals. One important design tradeoff may be quantified by noting that the sensitivity and complementary sensitivity functions satisfy the identity

$$S(j\omega) + T(j\omega) \equiv 1 \quad . \tag{2.3.3}$$

From this identity it follows that $|S(j\omega)|$ and $|T(j\omega)|$ cannot both be very small at the same frequency. Hence, at each frequency, there exists a tradeoff between those feedback properties such as sensitivity reduction and disturbance response which are quantified by $|S(j\omega)|$ and those properties such as noise response and robustness to high frequency uncertainty quantified by $|T(j\omega)|$.

In applications it often happens that levels of uncertainty and sensor noise become large at high frequencies, while disturbance rejection and sensitivity reduction are generally desired over a lower frequency range. Hence the tradeoff imposed by (2.3.3) is generally performed by requiring $M_S(\omega)$ to be small at low frequencies, $M_T(\omega)$ to be small at high frequencies, and neither $M_S(\omega)$ nor $M_T(\omega)$ to be excessively large at any frequency.

Recalling the terminology introduced in Section 2.1, we see that the tradeoff imposed by (2.3.3) is *algebraic*, since it occurs between system properties at the same frequency. An algebraic tradeoff also exists between $|S(j\omega)|$ and the response of the plant input to noise and disturbances. This tradeoff is present since, in practical situations, it is necessary to limit the magnitude of the plant input signal due to constraints imposed by the plant actuators. From (2.2.10), however, it follows that

$$\begin{aligned} u_d(s) &= -\, S(s)\, F(s)\, d(s) \\ &= -\, [1 - S(s)]P^{-1}(s)\, d(s) \quad , \end{aligned} \tag{2.3.4}$$

and, if $|S(j\omega)| \ll 1$, then

$$|u_d(j\omega)| \approx |P^{-1}(j\omega)|\ |d(j\omega)| \quad . \tag{2.3.5}$$

Hence, if sensitivity is small at a frequency for which plant gain is very small, then the response of the plant input to disturbances (and also noise) will be very large. The need to limit the size of this signal may therefore limit the frequency range over which sensitivity reduction can be obtained and the compensator gain can be allowed to exceed that of the plant. Equation (2.3.4) can be used to define an appropriate upper bound on $|T(j\omega)|$ that, if satisfied, will prevent response of the plant input to disturbances from being excessively large.

2.4: Relations Between Open and Closed Loop Properties of Scalar Feedback Systems

Classical "loop-shaping" design methods proceeded by directly manipulating open loop gain and phase (using, for example, lead and lag filters) to indirectly alter the feedback properties of the system. As is well-known, these methods are quite effective in coping with the type of design problems for which they were developed. Clearly, one reason for the success of these methods is that, for a scalar system, open loop gain and phase can be readily related to feedback properties. Indeed, the following relations are well-known (e.g.[Hor63, DoS81]) and can readily be deduced from (2.2.2-3):

$$|L(j\omega)| \gg 1 \iff \begin{array}{c} |S(j\omega)| \ll 1 \\ \text{and} \\ T(j\omega) \approx 1 \end{array} \tag{2.4.1}$$

and

$$|L(j\omega)| \ll 1 \iff \begin{array}{c} S(j\omega) \approx 1 \\ \text{and} \\ |T(j\omega)| \ll 1 \end{array} \tag{2.4.2}$$

At frequencies for which open loop gain is approximately unity, feedback properties depend critically upon the value of open loop phase:

$$\begin{array}{c} |L(j\omega)| \approx 1 \\ \text{and} \\ \measuredangle L(j\omega) \approx \pm 180^\circ \end{array} \iff \begin{array}{c} |S(j\omega)| \gg 1 \\ \text{and} \\ |T(j\omega)| \gg 1 \end{array} \tag{2.4.3}$$

Approximations (2.4.1-3) yield the following rules of thumb useful in design. First, large loop gain yields small sensitivity and good disturbance rejection properties, although noise appears directly in the system output. Second, small loop gain is required for small noise response and for robustness against large multiplicative uncertainty (2.2.14). Finally, at frequencies near gain crossover ($|L(j\omega)| = 1$), the phase of the system must remain bounded sufficiently far away from ±180° to provide an adequate stability margin and to prevent amplifying disturbances and noise.

It is also possible to relate open loop gain to the magnitude of the plant input. From (2.4.1) and (2.3.4), it follows that

$$|L(j\omega)| \gg 1 \quad \Leftrightarrow \quad |S(j\omega)\, F(j\omega)| \approx |P^{-1}(j\omega)| \quad . \tag{2.4.4}$$

Hence, requiring loop gain to be large at frequencies for which the plant gain is small may lead to unacceptably large response of the plant input to noise and disturbances[Hor63].

Recall our discussion of the sensitivity and complementary sensitivity specifications (2.3.1-2). From that discussion and the approximations (2.4.1-3), it follows [DoS81] that corresponding specifications upon open loop gain and phase might appear as in Figure 2.4.1. These specifications reflect the fact that loop gains are *desired* to be large at low frequencies for disturbance rejection and sensitivity reduction and are *required* to be small at high frequencies to provide stability robustness. At intermediate frequencies the phase of the system must remain bounded away from ±180° to prevent excessively large values of $|S|$ and $|T|$. In order that the benefits of feedback be obtained over as large a frequency range as possible, it is also desirable that ω_L should be close to ω_H. Of course, gain and phase must also satisfy the encirclement condition dictated by Nyquist's stability criteria.

Implicit in our construction of the design specification shown in Figure 2.4.1 is the assumption that open loop gain and phase can be independently manipulated. However, gain and phase are *not* mutually independent, and design tradeoffs must be taken into account when constructing specifications such as that in Figure 2.4.1.

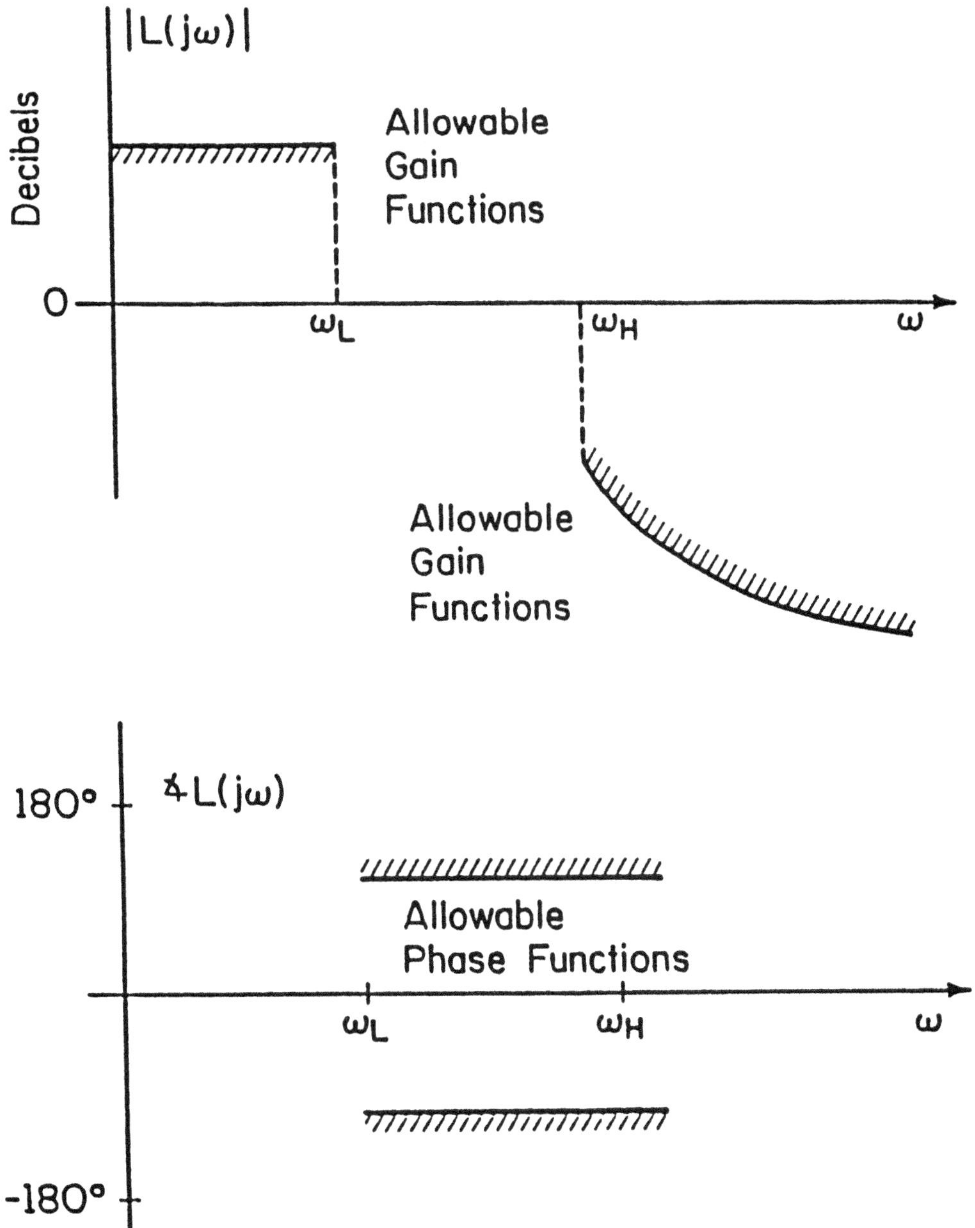

Figure 2.4.1: Gain and Phase Specifications

2.5: The Bode Gain-Phase Relation

We have shown in the previous section that design specifications can be stated in terms of the gain and phase of the open loop transfer function. For a large class of transfer functions, however, gain and phase are *not* mutually independent and the value of one is completely determined once the other is specified. There are many ways to state this relationship precisely; the one most useful for our purposes was derived by Bode[Bod45]. This relation has been used by many authors (c.f.[Hor63, DoS81, FPE86]) to analyze the implications that the gain-phase relation has upon feedback design.

Theorem 2.5.1 (Bode Gain-Phase Relation): Assume that $L(s)$ is a proper rational[1] function with real coefficients and with no poles or zeros in the closed right half plane. Then, at each frequency $s = j\omega_0$, the following integral relation must hold:

$$\angle L(j\omega_0) - \angle L(0) = \frac{1}{\pi} \int_{-\infty}^{\infty} \frac{d \log |L|}{d\nu} \left[\log \coth \frac{|\nu|}{2} \right] d\nu \quad , \tag{2.5.1}$$

where

$$\nu = \log(\omega/\omega_0) \tag{2.5.2}$$

■

Theorem 2.5.1 states conditions under which knowledge of open loop gain along the $j\omega$-axis suffices to determine open loop phase to within a factor of $\pm\pi$. Hence, for stable systems with no right half plane zeros, gain and phase cannot be manipulated independently in design.

The presence of the weighting function $\log \coth \frac{|\nu|}{2} = \log \left| \frac{\omega + \omega_0}{\omega - \omega_0} \right|$ shows that the dependence of $\angle L(j\omega_0)$ upon the rate of gain decrease at frequency ω diminishes rapidly as

[1]This assumption is made for simplicity of exposition; for a more general result, see[Bod45].

the distance between ω and ω_0 increases (Figure 2.5.1). Hence this integral supports a rule of thumb stating that a $20N$ db/decade rate of gain decrease in the vicinity of frequency ω_0 implies that $\sphericalangle L(j\omega_0) \approx -90N°$. It should be noted that this rule of thumb fails, for example, when the gain is piecewise constant. Indeed, Zadeh and Desoer ([ZaD63], p.432) present an example for which the value of phase is determined by an abrupt change in gain many decades away. Nevertheless, it seems that transfer functions encountered frequently in practice are sufficiently well-behaved that the value of phase at a frequency is largely determined by that of the gain over a decade-wide interval centered at the frequency of interest[FPE86].

The Bode gain-phase relation may be used to assess whether a design specification of the type shown in Figure 2.4.1 is achievable. Since a $20N$ db/decade rate of gain decrease in the vicinity of crossover implies that phase at crossover is roughly $-90N°$, it follows that the rate of gain decrease cannot be much greater than 20 db/decade if the Nyquist stability criterion is to be satisfied and if an acceptable phase margin is to be maintained. One implication of this fact is that the frequency ω_L in Figure 2.4.1 cannot be too close to the frequency ω_H. Hence the frequency range over which loop gain can be large to obtain sensitivity reduction is limited by the need to insure stability robustness against uncertainty at higher frequencies, and to maintain reasonable feedback properties near crossover. As discussed in [Hor63] and[DoS81], relaxing the assumption that $L(s)$ has no right half plane poles or zeros does not lessen the severity of this tradeoff; indeed, the tradeoff only becomes more difficult to accomplish. If one is willing to accept a system which is only *conditionally* stable, Horowitz [Hor63] shows that larger values of low frequency gain may be obtained.

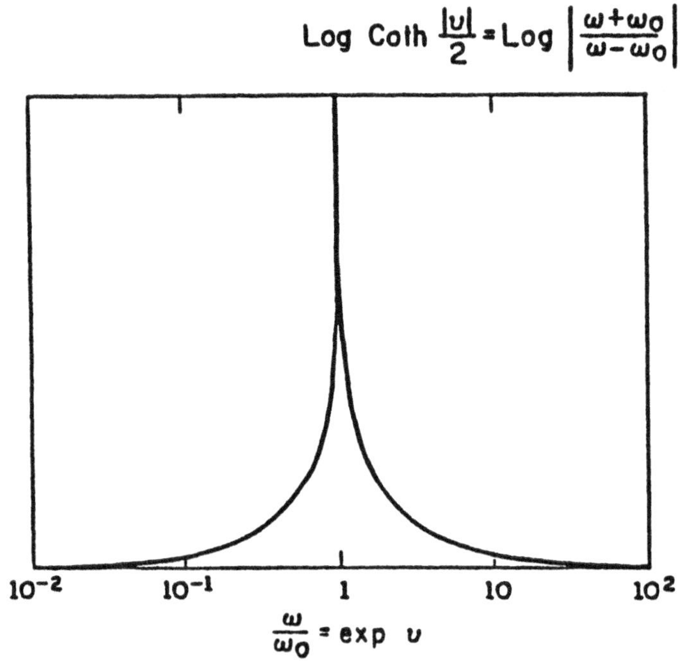

Figure 2.5.1: Weighting Function in Gain-Phase Integral

2.6: Summary

In this chapter we have briefly reviewed some important results from classical feedback theory; our purpose has been to identify aspects of that theory meriting extension to multivariable systems. It is convenient to divide the concepts needing extension into three groups.

First, it is obviously important to be able to evaluate the quality of a feedback design. The feedback properties discussed in this chapter can be readily evaluated, since they are directly expressed by the magnitude of certain closed loop transfer functions along the $j\omega$-axis. In Chapter 5 we shall see that closed loop transfer functions are also available with which to evaluate feedback properties of a multivariable system. Among these transfer functions are multivariable versions of the sensitivity and complementary sensitivity functions, as well as certain additional functions that have been developed to evaluate properties of multivariable systems having no scalar analogue.

Second, in any design methodology it is important to be able to relate the parameters available for adjustment by the designer to the feedback properties of the system. In classical loop-shaping design methodology, these parameters were the gain and phase of the open loop transfer function. The notion of gain can be extended to multivariable systems in a straightforward manner, although the notion of phase is more problematic. In Chapter 5 we shall discuss special cases where the rules of thumb relating open loop gain to feedback properties extend easily to multivariable systems. Chapter 6 will discuss the possible meaning of phase in multivariable systems and, in Chapter 7, we shall present additional rules of thumb for relating properties of the open loop transfer function to the feedback properties of a multivariable system.

The third fundamentally important class of results consists of the various algebraic and analytic constraints that lead to tradeoffs which must be performed in feedback design. As we shall discuss in Chapter 5, algebraic tradeoffs are still present; however, they now must be performed between system properties, in each *direction* at a given frequency. Analytic tradeoffs are also present, although they are quite difficult to quantify. Chapters 8-12 of this monograph will be devoted to developing mathematical tools needed to discuss tradeoffs between feedback properties of a multivariable system in different frequency ranges.

CHAPTER 3

TRADEOFFS IN THE DESIGN OF SCALAR LINEAR TIME-INVARIANT FEEDBACK SYSTEMS

3.1. Introduction

In the preceding chapter we saw that the sensitivity and complementary sensitivity functions directly express many important properties of a feedback design. For example, the magnitude of the sensitivity function evaluated along the $j\omega$-axis quantifies disturbance response properties of the system, while the magnitude of the complementary sensitivity function quantifies response to sensor noise. Since these functions are so important, we are motivated to state design specifications directly in terms of bounds upon their magnitude on the closed loop transfer function.

Obviously, one cannot construct design specifications such as the bounds on the closed loop transfer functions naively; indeed, one must take care that the specifications are *achievable*. If not, one must have available the insights necessary to make tradeoffs between conflicting design objectives. Several well-known limitations on achievable system properties and the resulting tradeoffs were reviewed in Chapter 2. Our purpose in the present chapter is to discuss some additional limitations and tradeoffs which have only recently been quantified. First, we shall briefly discuss the *sources* of the design tradeoffs of interest in this monograph.

It is possible to divide the sources of design limitations and tradeoffs into three (somewhat loosely defined) classes, according to their source. The first of these arises from the structure of a feedback loop itself. As an example, consider the tradeoff between the response of the system output to disturbances and to noise. Clearly, in order to use feedback to reduce disturbance response, it is necessary to measure the output of the system and therefore to become susceptible to the effects of sensor noise. Although this tradeoff is perhaps intuitively obvious, it is reassuring that it may be quantified precisely.

The second source of design limitations is the requirement that the feedback system be realizable. Since we are dealing only with systems that are adequately described by linear time-invariant models, the realizability requirement may be expressed using the transfer function representation of the model. As an example, consider the Bode gain-phase relation discussed in Section 2.5. The design tradeoff described by this relation is typical of those quantified by complex variable theory in that it manifests itself as a so-called *analytic* tradeoff between feedback properties in *different* frequency ranges. We shall discuss additional analytic tradeoffs in the remainder of this chapter.

Finally, design limitations also arise from properties of the plant which is to be controlled. It turns out that plants with properties such as right half plane zeros,[2] unstable poles, and time delays are inherently difficult to control. The degree of difficulty generally depends upon the design specifications imposed upon the system; if these specifications are too stringent, then it may be impossible to design a compensator so that they are satisfied. In that case, one must either relax the specifications, if possible, or else alter the structure of the plant. In either event, it is clearly important to develop analysis methods which yield both quantitative statements of the limitations inherent in a given plant as well as qualitative insights into how these limitations affect feedback properties.

Our purpose in Section 3.2 is to show that nonminimum phase zeros of the open loop transfer function (which must include those of the plant) impose an *analytic* tradeoff upon the sensitivity function of the system. Requiring sensitivity to be very small over one frequency range *necessarily* implies that sensitivity is large at other frequencies. An explicit statement of

---

[2]For reasons which shall become evident in Section 3.2, these are also referred to as *nonminimum phase* zeros.

this fact was first presented by Francis and Zames[FrZ84]. Insight into the limitations imposed by a nonminimum phase zero are also available using classical analysis techniques[Hor63]. The results we shall present have the double advantage that they are expressed directly in terms of the sensitivity function evaluated along the $j\omega$-axis and that they yield insight into the relative degree of difficulty caused by zeros in different locations.

In Section 3.3 we shall show that unstable open loop poles impose an analytic tradeoff upon the complementary sensitivity function. This tradeoff is essentially dual to that imposed upon the sensitivity function by nonminimum phase zeros.

A classical result due to Bode[Bod45], the Bode *sensitivity integral*, shows that a tradeoff exists between sensitivity properties at different frequencies whenever the open loop transfer function has at least two more poles than zeros. In Section 3.4, we shall extend this result to show that this tradeoff is more costly for systems which are open loop unstable. We shall also show that imposing a realistic bandwidth constraint at high frequencies requires a non-trivial sensitivity tradeoff to be performed in a lower frequency range. Finally, in Section 3.5, we present a summary and conclusions.

3.2. Nonminimum Phase Zeros and the Sensitivity Function

In Section 2.3 we saw that design specifications are often stated in terms of frequency dependent bounds on the magnitude of the sensitivity function: $|S(j\omega)| \le M_S(\omega), \ \forall \ \omega$. Small values of sensitivity are desirable to obtain the benefits of disturbance rejection and sensitivity reduction. In particular, at frequencies for which sensitivity is less than one, the disturbance response of the feedback system is smaller than that of an open loop design. On the other hand, at frequencies for which $|S(j\omega)| > 1$, use of feedback actually increases the response of the system to disturbances. We shall show in this section that when the plant is nonminimum phase, there is a tradeoff between these two types of behavior. Specifically, when the plant is nonminimum phase, requiring sensitivity to be less than one over some frequency range implies that sensitivity is necessarily greater than one at other frequencies.

It is interesting to note that this sensitivity tradeoff was stated explicitly only rather recently ([FrZ84, FrL85a]). However, nonminimum phase zeros have long been known to cause difficulty in design. Hence, before proceeding to derive a precise statement of the sensitivity tradeoff, we shall first discuss how the tradeoff may be seen qualitatively using classical analysis methods.

The magnitude of the sensitivity function of a scalar feedback system can be obtained easily using a Nyquist plot of $L(j\omega)$. Indeed, since $S(j\omega) = 1/[1+L(j\omega)]$, the magnitude of the sensitivity function is just the reciprocal of the distance from the Nyquist plot to the critical point (Figure 3.2.1). In particular, sensitivity is less than one at frequencies for which $L(j\omega)$ is outside the unit circle centered at the critical point. Sensitivity is greater than one at frequencies for which $L(j\omega)$ is inside this unit circle. We shall use this graphical analysis to investigate the effect that nonminimum phase zeros have upon the sensitivity function.

Suppose that we are presented with a plant possessing zeros in the open right half plane. Then the internal stability requirement dictates that these zeros also appear, with at least the same multiplicity, in the open loop transfer function $L(s) = P(s)F(s)$. Let the set of all open right half plane zeros of $L(s)$ (including any present in the compensator) be denoted by

$$\mathbf{Z} \triangleq \{z_i \, ; i = 1, \ldots, N_z\} \quad , \tag{3.2.1}$$

where each zero is counted according to its multiplicity. Defining the Blaschke product

$$B_z(s) \triangleq \prod_{i=1}^{N_z} \frac{z_i - s}{\bar{z}_i + s} \quad , \tag{3.2.2}$$

$$|S(j\omega)| = \left|\frac{1}{1+L(j\omega)}\right|$$

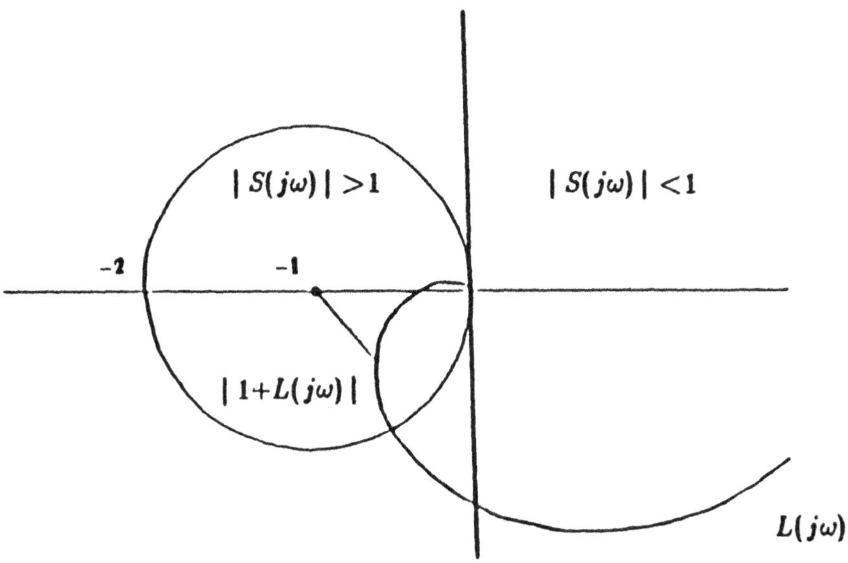

Figure 3.2.1: Sensitivity and the Nyquist Plot

we can then factor the open loop transfer function into the form

$$L(s) = L_m(s) \prod_{i=1}^{N_z} \frac{z_i - s}{\bar{z}_i + s} \tag{3.2.3}$$

where $L_m(s)$ has no zeros in the open right half plane. Note that

$$|L(j\omega)| = |L_m(j\omega)| \qquad \forall\ \omega \tag{3.2.4}$$

and

$$\measuredangle \frac{z_i - j\omega}{\bar{z}_i + j\omega} \to -180° \qquad \text{as } \omega \to \infty \quad . \tag{3.2.5}$$

These facts show that open right half plane zeros contribute additional phase lag without changing the gain of the system (hence the term "nonminimum phase zero"). The effect that this additional lag has upon feedback properties can best be illustrated using a simple example.

Consider the nonminimum phase plant $P(s) = \frac{1}{s+1} \cdot \frac{1-s}{1+s}$ and its nonminimum phase counterpart $P_m(s) = \frac{1}{s+1}$. From Figure 3.2.2, we see that the additional phase lag contributed by the zero at $s = 1$ causes the Nyquist plot to penetrate the unit circle and sensitivity to be larger than one. Experiments with various compensation schemes reveal that using large loop gain over some frequency range to obtain small sensitivity in that range tends to cause sensitivity to be large at other frequencies.

The preceding observations motivate us to conjecture that a tradeoff between sensitivity properties in different frequency ranges is a design limitation *inherent* in a system with a nonminimum phase plant. We therefore seek a precise mathematical statement of this limitation, if it indeed exists, as well as additional insights into the qualitative nature of the tradeoff. For example, intuition suggests that if sensitivity reduction is required *only* over a frequency range in which the zero contributes little additional phase lag, then the detrimental effect of the zero

$$P(s) = \frac{1}{s+1} \cdot \frac{1-s}{1+s} \qquad \tilde{P}(s) = \frac{1}{s+1}$$

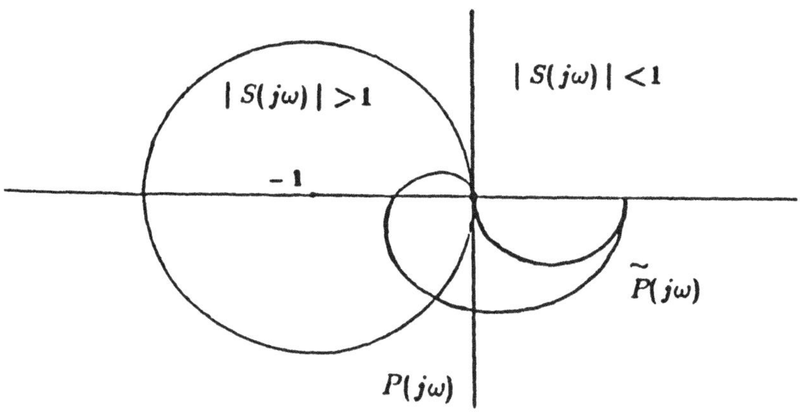

Figure 3.2.2: Additional Phase Lag Contributed by a Nonminimum Phase Zero

should be negligible.

We shall assume that the open loop transfer function can be factored as

$$L(s) = L_0(s)B_z(s)B_p^{-1}(s)e^{-s\tau} \quad , \tag{3.2.6}$$

where $\tau \geq 0$ represents a possible time delay, $L_0(s)$ is a proper rational[3] function with no poles or zeros in the open right plane, and $B_z(s)$ is the Blaschke product (3.2.2) containing the open right half plane zeros of the plant plus those of the compensator. The Blaschke product

$$B_p(s) = \prod_{i=1}^{N_p} \frac{p_i - s}{\overline{p}_i + s} \tag{3.2.7}$$

contains all the poles of both plant and compensator in the open right half plane, again counted according to multiplicity. We emphasize once more that internal stability requirements dictate that all right half plane poles and zeros of the plant *must* appear with at least the same multiplicity in $L(s)$ and hence cannot be cancelled by right half plane zeros and poles of the compensator.

One constraint that right half plane zeros impose upon the sensitivity function is immediately obvious from the definition $S(s) = 1/[1+L(s)]$. Suppose that $L(s)$ has a zero with multiplicity m at $s = z$. It follows that

$$S(z) = 1$$

and (3.2.8)

$$\frac{d^i}{ds^i} S\Big|_{s=z} = 0 \ , \quad i = 1, \ldots, m-1 \ .$$

Poles of $L(s)$ also constrain the sensitivity function. If $L(s)$ has a pole with multiplicity n at

---

[3]The assumption that $L_0(s)$ is rational can be relaxed somewhat; see[FrL85a].

$s = p$, then

$$S(p) = 0$$

and (3.2.9)

$$\frac{d^i}{ds^i} S \Big|_{s=p} = 0 \ , \quad i = 1, \ldots, n-1 \ .$$

From (3.2.8-9), it is clear that if the plant, and thus $L(s)$, has zeros or poles at points of the open right half plane, then the value of the sensitivity function is constrained at those points. Naturally, we are more concerned about the value of sensitivity along the $j\omega$-axis, where the design specifications are imposed and the conjectured tradeoff must take place. Using complex variable theory, we shall now show that the constraint (3.2.8) on sensitivity at an open right half plane zero has an equivalent statement in terms of sensitivity along the $j\omega$-axis.

The mathematical tool we need is the Poisson integral formula ([LeR70, Con78]). Let $f(s)$ be analytic in the closed right half plane and define

$$M(R) = \max_{\theta} |f(Re^{j\theta})| \ , \quad \theta \in [-\frac{\pi}{2}, \frac{\pi}{2}] \ . \tag{3.2.10}$$

Suppose that

$$\lim_{R \to \infty} M(R)/R = 0 \ . \tag{3.2.11}$$

Then the value of $f(s)$ at any point $s = x + jy$ in the open right half plane can be recovered from the values of $f(j\omega)$ via the integral relation

$$f(s) = \frac{1}{\pi} \int_{-\infty}^{\infty} f(j\omega) \frac{x}{x^2 + (y-\omega)^2} \, d\omega \ . \tag{3.2.12}$$

Alternately, if $f(s)$ satisfies the appropriate conditions, then constraining this function to take a certain value at a point of the open right half plane yields an integral constraint upon $f(j\omega)$.

It is this latter interpretation which shall be of use to us.

Suppose that $L(s)$ has the form (3.2.6) and that the sensitivity function $S(s) = 1/[1+L(s)]$ is *stable*. We would like to apply (3.2.12) to a branch of the logarithm of the sensitivity function; however, an analytic branch of the log function cannot be defined if $S(s)$ has zeros in the open right half plane. Since these zeros are due solely to the open right half plane poles of $L(s)$, we can factor them out using the Blaschke product (3.2.7) and define

$$\tilde{S}(s) \triangleq S(s)B_p^{-1}(s) \quad . \tag{3.2.13}$$

The assumption that $S(s)$ is stable implies that $\tilde{S}(s)$ is stable also. Furthermore, by construction, $\tilde{S}(s)$ is nonzero in the open right half plane. Hence we can define a function $\log\tilde{S}(s)$ which is analytic in the open right half plane and such that $\exp[\log\tilde{S}(s)] = \tilde{S}(s)$. Furthermore

$$|\tilde{S}(j\omega)| = |S(j\omega)| \tag{3.2.14}$$

and, if $L(z) = 0$,

$$\tilde{S}(z) = B_p^{-1}(z) \quad . \tag{3.2.15}$$

Using these facts, Poisson's integral formula (3.2.12) may be manipulated to yield the following integral constraints on the sensitivity function.

Theorem 3.2.1: Suppose that the open loop transfer function $L(s)$ has a zero, $z = x + jy$, with $x > 0$. Assume that the associated feedback system is stable. Then the sensitivity function must satisfy

$$\int_0^\infty \log|S(j\omega)|\, W(z,\omega)d\omega = \pi\log|B_p^{-1}(z)| \quad , \tag{3.2.16}$$

where $W(z,\omega)$ is a weighting function. For a real zero, $z = x$,

$$W(x,\omega) = \frac{2x}{x^2+\omega^2} \qquad (3.2.17)$$

and, for a complex zero, $z = x + jy$,

$$W(z,\omega) = \frac{x}{x^2+(y-\omega)^2} + \frac{x}{x^2+(y+\omega)^2} \quad . \qquad (3.2.18)$$

In addition, for a complex zero, the sensitivity function must satisfy

$$\int_0^\infty \log|S(j\omega)|\ V(z,\omega)d\omega = \pi \measuredangle B_p^{-1}(z) \qquad (3.2.19)$$

where

$$V(z,\omega) = \frac{\omega - y}{x^2 + (y-\omega)^2} - \frac{\omega + y}{x^2 + (y+\omega)^2} \qquad (3.2.20)$$

■

Proof: Equations (3.2.16-18) follow easily from Theorem 1 in [FrL85a] using conjugate symmetry: $\tilde{S}(\bar{s}) = \overline{\tilde{S}(s)}$. Equations (3.2.19-20) follow from a result due to Stein [StA84].

■

A number of remarks about this theorem are in order. First, as discussed in[FrL85a], the integral relations are valid even if $S(s)$ has zeros (or poles) on the $j\omega$-axis. Second, if z has multiplicity $m > 1$, then the additional constraints (3.2.8) on the first $m - 1$ derivatives of $\log S(s)$ also have equivalent statements as integral relations[FrL85a].

For our purposes, the integral relation of greatest interest is (3.2.16), since it is this relation which implies the existence of the sensitivity tradeoff. To see this, note first that the weighting function satisfies $W(z,\omega) > 0,\ \forall\ \omega$, and the Blaschke product satisfies $\log|B_p^{-1}(z)| \geq 0$. Using these facts, it follows easily from (3.2.16) that requiring $\log|S(j\omega)| < 0$ over some frequency range implies that $\log|S(j\omega)| > 0$ at other frequencies. Hence, if the plant is nonminimum phase, one cannot use feedback to obtain the benefits of sensitivity reduction over one frequency range unless one is willing to pay the attendant price in terms of increased sensitivity elsewhere.

Theorem 3.2.1 verifies our conjecture that a sensitivity tradeoff is present whenever a system is nonminimum phase. Recall we also conjectured that the severity of the tradeoff is a function of the phase lag contributed by the zero at frequencies for which sensitivity reduction is desired. Using the weighting function $W(z,\omega)$ (3.2.17-18), we shall now show that this conjecture is also true.

Consider first the case of a real zero. From (3.2.2), we see that, as a function of frequency, the additional phase lag contributed by this zero is

$$\theta(x,\omega) \triangleq \sphericalangle \frac{x - j\omega}{x + j\omega} \tag{3.2.21}$$

Noting that

$$\frac{d\theta(x,\omega)}{d\omega} = \frac{-2x}{x^2 + \omega^2} \tag{3.2.22}$$

it follows that the weighting function in (3.2.17) satisfies

$$W(x,\omega) = -\frac{d\theta(x,\omega)}{d\omega} \quad . \tag{3.2.23}$$

Hence the weighting function appearing in the sensitivity constraint is equal to (minus) the rate at which the phase lag due to the zero increases with frequency.

One can use the weighting function (3.2.23) to compute the weighted length of a frequency interval. Note that sensitivity reduction is typically required over a low frequency interval $\Omega \triangleq [0,\omega_1]$ and that the weighted length of such an interval equals

$$W(x,\Omega) \triangleq \int_0^{\omega_1} W(x,\omega)d\omega$$

$$= -\theta(x,\omega_1) \quad . \tag{3.2.24}$$

Hence the weighted length of the interval is equal to (minus) the phase lag contributed by the zero at the upper endpoint of the interval. It follows that, as $\omega_1 \to \infty$, the weighted length of the $j\omega$-axis equals π. We shall investigate the effect that the phase lag has upon the sensitivity tradeoff momentarily. First, we note that the weighting function (3.2.18) due to a complex zero is equal to $(-1/2)$ times the sum of the additional phase lag contributed by the zero and by its complex conjugate:

$$W(z,\omega) = -\frac{1}{2}\left[\frac{d\theta(z,\omega)}{d\omega} + \frac{d\theta(\bar{z},\omega)}{d\omega}\right] \quad . \tag{3.2.25}$$

Hence the weighted length of the frequency interval $\Omega = [0,\omega_1]$ is

$$W(z,\omega) = -\frac{1}{2}\left[\theta(z,\omega_1) + \theta(\bar{z},\omega_1)\right] \quad . \tag{3.2.26}$$

As we have already remarked, the integral constraint (3.2.16) implies that a tradeoff exists between sensitivity reduction and sensitivity increase in different frequency ranges. An interesting interpretation of this tradeoff is available using the weighting function. Suppose first that $L(s)$ has no poles in the open right half plane. Then the integral constraint becomes

$$\int_0^{\infty} \log|S(j\omega)|\,W(x,\omega)d\omega = 0 \tag{3.2.27}$$

and we see that the weighted area of sensitivity increase must *equal* the weighted area of sensitivity reduction. Since the weighted length of the $j\omega$-axis is *finite*, it follows that the amount by which sensitivity must exceed one at higher frequencies cannot be made arbitrarily small.

If the open loop system has poles in the open right half plane, then the weighted area of sensitivity increase must *exceed* that of sensitivity reduction. In particular, since

$$\log|B_p^{-1}(z)| = \sum_{i=1}^{N_p} \log\left|\frac{\bar{p}_i + z}{p_i - z}\right| \tag{3.2.28}$$

it follows (unsurprisingly) that systems with approximate pole-zero cancellations in the open right half plane can have especially bad sensitivity properties.

We can use the integral constraint (3.2.16) to obtain some simple lower bounds on the size of the peak in sensitivity accompanying a given level of sensitivity reduction over a low frequency interval. Bounds of this type were first discussed by Francis and Zames ([FrZ84], Theorem 3). The results we present will show how the relative location of the zero to the interval of sensitivity reduction influences the size of the peak in sensitivity outside that interval.

Suppose that we require the sensitivity function to satisfy the upper bound

$$|S(j\omega)| \le \alpha < 1 \quad , \forall\ \omega \in \Omega \ , \tag{3.2.29}$$

where $\Omega = [0, \omega_1]$ is a low frequency interval of interest. Define the infinity norm of the sensitivity function:

$$\|S\|_\infty = \sup_{\omega \ge 0} |S(j\omega)| \ . \tag{3.2.30}$$

Assuming that the upper bound (3.2.29) is satisfied, the integral (3.2.16) may be used to compute a lower bound on $\|S\|_\infty$ for each nonminimum phase zero of $L(s)$.

Corollary 3.2.2: Suppose that the conditions in Theorem 3.2.1 are satisfied and that the sensitivity function is bounded as in (3.2.29). Then the following lower bound must be satisfied at each nonminimum phase zero of $L(s)$:

$$\|S\|_\infty \geq (1/\alpha)^{\frac{W(z,\Omega)}{\pi - W(z,\Omega)}} \, |B_p^{-1}(z)|^{\frac{\pi}{\pi - W(z,\Omega)}} \quad . \tag{3.2.31}$$

■

Proof: From (3.2.16),

$$\pi \log |B_p^{-1}(z)| = \int_0^{\omega_1} \log |S(j\omega)| W(z,\omega) d\omega + \int_{\omega_1}^{\infty} \log |S(j\omega)| W(z,\omega) d\omega$$

(3.2.32)

$$\leq \log\alpha \cdot W(z,\Omega) + \log \|S\|_\infty (\pi - W(z,\Omega)) \quad .$$

from which the result follows.

■

The bound (3.2.31) shows that requiring sensitivity to be very small over the interval $(0,\omega_1)$ implies that there necessarily exists a large peak in sensitivity outside this interval. Furthermore, the smallest possible size of this peak will become larger if the open loop system has unstable poles near any zero.

The size of the sensitivity peak also depends upon the location of the interval $(0,\omega_1)$ relative to the zero. To see why this is so, assume for simplicity that the system is open loop stable and that the zero is real. Hence

$$\|S\|_\infty \geq (1/\alpha)^{\frac{W(x,\Omega)}{\pi - W(x,\Omega)}} \quad . \tag{3.2.33}$$

Recall that the weighted length of the interval $\Omega = (0,\omega_1)$ is just equal to (minus) the phase lag contributed by the zero at the upper endpoint of that interval. Since the zero eventually contributes 180° phase lag, it follows that as $\omega_1 \to \infty$, $W(x,\Omega) \to \pi$. Thus the exponent in (3.2.33) becomes unbounded and, since $\alpha < 1$, so does the peak in sensitivity. To summarize, requiring sensitivity to be small throughout a frequency range extending into the region where the non-minimum phase zero contributes a significant amount of phase lag implies that there will, of necessity, exist a large peak in sensitivity at higher frequencies. On the other hand, if the zero is located so that it contributes only a negligible amount of phase lag at frequencies for which sensitivity reduction is desired, then it does not impose a serious limitation upon sensitivity properties of the system. Analogous results hold, with appropriate modifications, for a complex zero.

Suppose now that the open loop system has poles in the open right half plane. It is interesting to note that, in this case, the bound (3.2.31) implies the existence of a peak in sensitivity even if no sensitivity reduction is present!

Recall next the rule of thumb (2.4.1) which shows that small sensitivity can be obtained only by requiring open loop gain to be large. In fact, it is easy to show that requiring $|S(j\omega)| \leq \alpha < 1$ implies that $|L(j\omega)| \geq 1/\alpha - 1$. From our analysis of (3.2.33), it follows that, to prevent poor feedback properties, open loop gain should not be large over a frequency interval extending into the region for which a nonminimum phase zero contributes significant phase lag. This observation substantiates a classical design rule of thumb stating that loop gain must be rolled off before the phase lag contributed by the zero becomes significant. We should note, however, that if one is willing and able to adopt some nonstandard design strategies, such as having multiple gain crossover frequencies, then [HoL84] it is possible to manipulate the design tradeoff imposed by a nonminimum phase zero to obtain some benefits of large loop gain at higher frequencies. One drawback of these strategies is that loop gain must be small, and hence the benefits of feedback must be lost, over an *intermediate* frequency range.

Finally, as is pointed out in [FrL85a], the bound (3.2.31) is *tight* when only one (real) nonminimum phase zero is present. Tight bounds in the case of multiple zeros may be obtained using the interesting results of O'Young and Francis[OYF85].

3.3. Unstable Poles and the Complementary Sensitivity Function

We shall show in this section that unstable poles impose constraints upon the complementary sensitivity function which, loosely speaking, are *dual* to those imposed upon the sensitivity function by nonminimum phase zeros. That such constraints exist might be conjectured from the constraints (3.2.9) and the identity $S(s) + T(s) \equiv 1$. Together, these equations show that if $L(s)$ has a pole with multiplicity n at $s = p$, then the complementary sensitivity function satisfies

$$T(p) = 1$$

and (3.3.1)

$$\frac{d^i}{di} T \bigg|_{s=p} = 0 \quad , \quad i = 1, \ldots, n-1 \quad .$$

Furthermore, if $L(s)$ has a zero with multiplicity m at $s = z$, then

$$T(z) = 0$$

and (3.3.2)

$$\frac{d^i}{ds^i} T \bigg|_{s=z} = 0 \quad , \quad i = 1, \ldots, m-1 \quad .$$

Our previous work on the sensitivity function, together with the fact that $T(s)$ is constrained to equal one at open right half plane poles of $L(s)$, suggests that Poisson's integral formula might be applied to derive a constraint on $|T(j\omega)|$ due to the presence of such poles. It should also be possible to motivate the presence of the integral constraint on $|T(j\omega)|$ using an argument based upon the *inverse* Nyquist plot [Ros74] and the fact that $|T(j\omega)| > 1$ whenever $L^{-1}(j\omega)$ is inside the unit circle centered at the critical point. However, we shall not pursue these details here.

As in the previous section, we shall assume that $L(s)$ has the form (3.2.6). Note this implies that $T(s)$ may be factored as

$$T(s) = \tilde{T}(s)B_z(s)e^{-s\tau} \quad , \tag{3.3.3}$$

where $B_z(s)$ is the Blaschke product (3.2.2) of open loop zeros and $e^{-s\tau}$ is the open loop time delay. In order to apply Poisson's integral formula, it is necessary that we work with the function $\tilde{T}(s)$, which is minimum phase by construction and is stable if $T(s)$ is stable. These facts allow us to define a function $\log\tilde{T}(s)$ which is analytic in the open right half plane and which satisfies $\exp[\log\tilde{T}(s)] = \tilde{T}(s)$. We also remove the time delay from $T(s)$ in order that condition (3.2.11) be satisfied. Since $B_z(s)$ and $e^{-s\tau}$ are both all pass with unit modulus, it follows that

$$|\tilde{T}(j\omega)| = |T(j\omega)| \tag{3.3.4}$$

and, if $L(s)$ has a pole at $s = p$,

$$\tilde{T}(p) = B_z^{-1}(p)e^{\tau p} \quad . \tag{3.3.5}$$

Using these facts, Poisson's integral formula may be applied to $\tilde{T}(s)$, yielding the following integral constraint on the complementary sensitivity function.

Theorem 3.3.1: Suppose that the open loop transfer function has a pole, $p = x + jy$, with $x > 0$. Assume that the associated feedback system is stable. Then the complementary sensitivity function must satisfy

$$\int_0^\infty \log|T(j\omega)|W(p,\omega)d\omega = \pi\log|B_z^{-1}(p)| + \pi x \tau \quad , \tag{3.3.6}$$

where $W(p,\omega)$ is a weighting function. For a real pole, $p = x$,

$$W(x,\omega) = \frac{2x}{x^2+\omega^2} \tag{3.3.7}$$

and, for a complex pole, $p = x + jy$

$$W(p,\omega) = \frac{x}{x^2 + (y - \omega)^2} + \frac{x}{x^2 + (y + \omega)^2} \quad . \tag{3.3.8}$$

In addition, for a complex pole, the complementary sensitivity function must satisfy

$$\int_0^\infty \log|T(j\omega)| V(p,\omega) d\omega = \pi \measuredangle B_z^{-1}(p) + \pi y \tau \tag{3.3.9}$$

where

$$V(p,\omega) = \frac{\omega - y}{x^2 + (y - \omega)^2} - \frac{\omega + y}{x^2 + (y + \omega)^2} \quad . \tag{3.3.10}$$

■

Proof: Equations (3.3.6-8) follow easily from Theorem 2 in [FrL85a] using conjugate symmetry: $\bar{T}(\bar{s}) = \overline{\bar{T}(s)}$. Equations (3.3.9-10) follow from a result due to Stein[StA84].

■

Remarks analogous to those following Theorem 1 apply to this result also. The integral relations are valid even if $T(s)$ has zeros on the $j\omega$-axis, and there are additional constraints on the derivative of $\log T(s)$ at poles with multiplicity greater than one.

The integral (3.3.6) shows that there exists a tradeoff between sensor noise response properties in different frequency ranges whenever the system is open loop unstable. Since $|T(j\omega)|$ is the reciprocal of the stability margin against multiplicative uncertainty, it follows that a tradeoff between stability robustness properties in different frequency ranges also exists. Using analysis methods similar to those in the preceding section, one can derive a lower bound on the peak in the complementary sensitivity function present whenever $|T(j\omega)|$ is required to be small over some frequency interval. One difference is that $|T(j\omega)|$ is generally required to be small over a high, rather than a low, frequency range.

It is interesting that time delays worsen the tradeoff upon sensor noise reduction imposed by unstable poles. This is plausible for the following reason. Use of feedback around an open-loop unstable system is necessary to achieve stability. Time delays, as well as non-minimum phase zeros, impede the processing of information around a feedback loop. Hence, it is reasonable to expect that design tradeoffs due to unstable poles are exacerbated when time delays and/or nonminimum phase zeros are present. This interpretation is substantiated by the fact that the term due to the time delay in (3.3.6) is proportional to both the length of the time delay and the distance from the unstable pole to the left half plane.

3.4. Bode's Sensitivity Integral and an Extension

The purpose of this section is to discuss the Bode sensitivity integral[Bod45, Hor63] alluded to in our discussion of analytic design tradeoffs in Section 2.3. This integral quantifies a tradeoff between sensitivity reduction and sensitivity increase which must be performed whenever the open loop transfer function has at least two more poles than zeros.

To motivate existence of the integral constraint, consider the open loop transfer function $L(s) = \frac{2}{(s+1)^2}$. As shown in Figure 3.4.1, there exists a frequency range over which the Nyquist plot of $L(j\omega)$ penetrates the unit circle and sensitivity is thus greater than one. In practice, the open loop transfer function will generally have *at least* two more poles than zeros [Hor63]. If $L(s)$ is stable then, using the gain-phase relation (2.5.1), it is straightforward to show that $L(j\omega)$ will asymptotically have phase lag at least $-180°$. Hence there will always exist a frequency range over which sensitivity is greater than one. This behavior may be quantified using a classical theorem due to Bode[Bod45].

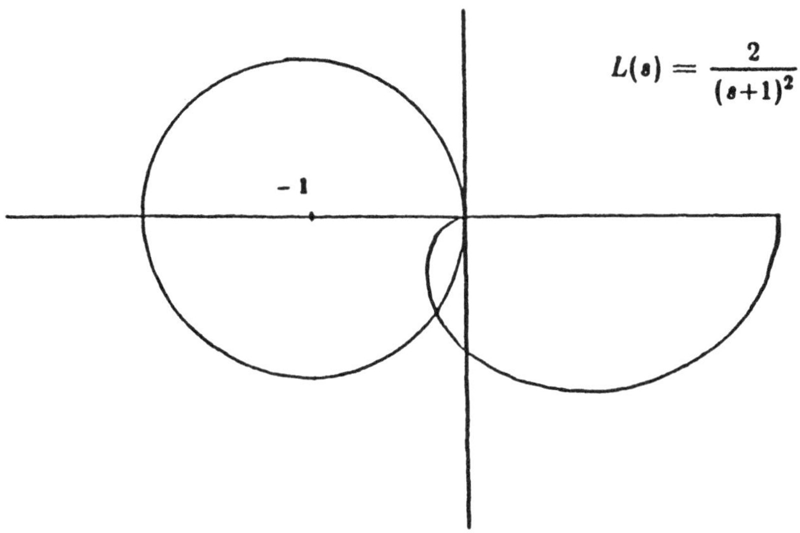

Figure 3.4.1: Effect of Two-Pole Rolloff Upon Nyquist Plot

Theorem 3.4.1 (Bode Sensitivity Integral) Suppose that the open loop transfer function $L(s)$ is rational[4] and has no poles in the open right half plane. If $L(s)$ has at least two[5] more poles then zeros, and if the associated feedback system is stable, then the sensitivity function must satisfy

$$\int_0^\infty \log|S(j\omega)|\,d\omega = 0 \quad . \tag{3.4.1}$$

■

This theorem shows that a tradeoff exists between sensitivity properties in different frequency ranges. Indeed, the area of sensitivity reduction must *equal* the area of sensitivity increase in units of decibels × (rad/sec). In this respect, the benefits and costs of feedback are balanced exactly. Before discussing the design implications of Theorem 3.4.1, we shall generalize the result to systems with unstable poles in the open right half plane.

Theorem 3.4.2[FrL85a]: Suppose that all assumptions in Theorem 3.4.1 are satisfied *except* that we allow $L(s)$ to have poles in the open right half plane. Let the set of these poles, counted according to multiplicity, be denoted $\{p_i \; ; \; i = 1, \ldots, N_p\}$. Then the sensitivity function must satisfy

$$\int_0^\infty \log|S(j\omega)|\,d\omega = \pi \sum_{i=1}^{N_p} \mathrm{Re}[p_i] \quad . \tag{3.4.2}$$

■

This extension of Bode's theorem shows that when the open loop system is unstable, then the area of sensitivity increase *exceeds* that of sensitivity reduction by an amount proportional to the distance from the unstable poles to the left half plane. A little reflection reveals that this

---

[4]This assumption can be relaxed. See, for example,[FrL85a].

[5]If the pole-zero excess is only one, then the integral is finite, but has a nonzero value[Hor63, KwS72].

additional sensitivity increase is plausible for the following reason. When the system is open loop unstable, then it is obviously necessary to use feedback to achieve closed loop stability, as well as to obtain sensitivity reduction. One might expect that this additional benefit of feedback would be accompanied by a certain cost, and the integral (3.4.2) substantiates this hypothesis. Alternately, we could interpret (3.4.2) as implying that the area of sensitivity reduction must be *less* than that of sensitivity increase, thus indicating that a portion of the open loop gain which could otherwise contribute to sensitivity reduction must instead be used to pull the unstable poles into the left half plane.

The integral (3.4.1-2) is conceptually interesting; however, in practice the tradeoff it quantifies does not impose a meaningful design limitation. Although it is true that requiring a large area of sensitivity reduction over a low-frequency interval implies that an equally large area of sensitivity increase must be present at higher frequencies, it does not follow that there must exist a peak in sensitivity which is bounded greater than one. Indeed, it is possible to achieve an arbitrarily large area of sensitivity increase by requiring $|S(j\omega)| = 1 + \delta$, $\forall\ \omega \in (\omega_1, \omega_2)$, where δ can be chosen arbitrarily small and the interval (ω_1, ω_2) is adjusted to be sufficiently large. By way of contrast, recall that this strategy cannot be applied to the tradeoff imposed by a nonminimum phase zero, since the effective length of the $j\omega$-axis in (3.2.16) is *finite*.

Our analysis in the preceding paragraph — where we concluded that the sensitivity integral imposes no meaningful design limitation — is in fact quite naive. The reason is that we neglected to consider the effect of limitations upon system *bandwidth* that are always present in a practical design. For example, as we saw in Section 2.4, it is necessary to decrease open loop gain at high frequencies in order to maintain stability robustness against large modelling errors due to unmodelled dynamics. Small open loop gain is also required to prevent sensor noise from appearing at the system output. Finally, requiring open loop gain to be large at a frequency for which plant gain is small may lead to unacceptably large response of the plant input to noise and disturbances. Hence the natural bandwidth of the plant also imposes a limitation upon open loop bandwidth.

One or more of the bandwidth constraints just cited is usually present in any practical design. It is reasonable, therefore, to assume that open loop gain must satisfy a frequency dependent bound of the form

$$|L(j\omega)| \leq \varepsilon \left[\frac{\omega_c}{\omega}\right]^{1+k} \quad , \quad \forall \quad \omega \geq \omega_c \quad , \tag{3.4.3}$$

where $\varepsilon < 1/2$ and $k > 0$. This bound imposes a constraint upon the rate at which loop gain rolls off, as well as the frequency at which roll-off commences and the level of gain at that frequency.

When a bandwidth constraint such as (3.4.3) is imposed, it is obviously not possible to require the sensitivity function to exceed one over an arbitrarily large frequency interval, as we suggested in our "naive" discussion of the sensitivity tradeoff. Indeed, when (3.4.3) is satisfied, there is an *upper* bound on the area of sensitivity increase which can be present at frequencies greater than ω_c.

Corollary 3.4.3: Suppose, in addition to the assumptions of Theorem 3.4.2, that $L(s)$ satisfies the bandwidth constraint (3.4.3). Then the tail of the sensitivity integral must satisfy

$$\left| \int_{\omega_c}^{\infty} \log|S(j\omega)| d\omega \right| \leq \frac{3\varepsilon\omega_c}{2k} \quad . \tag{3.4.4}$$

■

Proof: On page 12 of [Con78], it is shown that, if z is a complex number with $|z| < 1/2$, then $|\log(1+z)| \leq \frac{3}{2}|z|$. Using this fact, we have

$$
\begin{aligned}
\left| \int_{\omega_c}^{\infty} \log|S(j\omega)|\, d\omega \right| &\le \int_{\omega_c}^{\infty} \left| \log|S(j\omega)| \right| d\omega \\
&\le \int_{\omega_c}^{\infty} \left| \log S(j\omega) \right| d\omega \\
&= \int_{\omega_c}^{\infty} \left| \log[1+L(j\omega)] \right| d\omega \\
&\le \frac{3\varepsilon\omega_c^{1+k}}{2} \int_{0}^{\infty} \left[\frac{1}{\omega} \right]^{1+k} d\omega \\
&= \frac{3\varepsilon\omega_c^{1+k}}{2} \cdot \left[-\omega^{-k}/k \right] \Big|_{\omega_c}^{\infty}
\end{aligned}
$$

■

The bound (3.4.4) implies that the sensitivity tradeoff imposed by the integral (3.4.2) must be accomplished primarily over a finite frequency interval. As a consequence, the amount by which $|S(j\omega)|$ must exceed one cannot be arbitrarily small. Indeed, suppose that the sensitivity function is required to satisfy the upper bound

$$
|S(j\omega)| \le \alpha < 1 \ , \ \forall \ \ \omega \le \omega_1 < \omega_c \tag{3.4.5}
$$

If the bandwidth constraint (3.4.3) and the sensitivity bound (3.4.5) are *both* satisfied, then the integral constraint (3.4.2) may be manipulated readily to show that

$$
\sup_{\omega \in (\omega_1, \omega_c)} \log|S(j\omega)| \ge \frac{1}{(\omega_c - \omega_1)} \left\{ \pi \sum_{i=1}^{N_p} \mathrm{Re}[p_i] + \omega_1 \cdot \log(1/\alpha) - \frac{3\varepsilon\omega_c}{2k} \right\} \tag{3.4.6}
$$

The bound (3.4.6) shows that increasing the area of low frequency sensitivity reduction by requiring α to be very small and/or ω_1 to be very close to ω_c will *necessarily* cause a large peak in sensitivity at frequencies between ω_1 and ω_c. Hence the integral constraint (3.4.6) *together* with the bandwidth constraint (3.4.3) imposes a tradeoff between sensitivity reduction and sensitivity increase which must be accounted for in design.

Finally, we note that analogous results can be obtained if the bandwidth constraint is imposed upon the complementary sensitivity function rather than on the open loop transfer function. Motivation for imposing the constraint upon $T(s)$ rather than upon $L(s)$ stems from the fact that $T(s)$ directly expresses the feedback properties of sensor noise response and stability robustness, while $L(s)$ does so only indirectly.

3.5. Summary and Conclusions

We have discussed three potential limitations upon the ability of a feedback system to meet desired design specifications. Each of these limitations manifests itself as an analytic tradeoff between system properties in different frequency ranges. To summarize, nonminimum phase zeros impose a tradeoff upon the sensitivity function and thus upon the feedback properties which it describes, while unstable poles do the same for the complementary sensitivity function. It is worth pointing out that, in the integrals (3.2.16) and (3.3.6), no distinction is made between right half plane poles and zeros of the plant and those of the compensator. From a practical standpoint, however, there is a fundamental difference in that we are often *given* the plant, and required to *choose* a compensator. Hence, we may have no control over the right half plane *plant* poles and zeros, and thus over the associated design limitations, while we do have control over the poles and zero of the compensator[6] Hence, presumably, one should be able to select a compensator which does not itself cause significant design difficulty. On the other hand, the design difficulties inherent in the plant may prevent the original design specification from being satisfied. In that case, the insight furnished by the weighting function in (3.2.16) and (3.3.6) should be useful in trading off design goals so that an achievable

---

[6]An exception being certain cases where plants which are both unstable *and* nonminimum phase cannot be stabilized by a stable compensator. For a characterization of such plants, see [Vid85] and the references contained therein.

specification is obtained. Of course it may well be that no satisfactory design specification can be achieved. In this case, the integral constraints should aid in modification of the plant so that it becomes possible to obtain reasonable feedback properties. Indeed, one advantage of having a precise understanding of the design limitations is that this information could be used, at an *earlier* step in the engineering process, to *construct* a plant which is not inherently too difficult to reliably control.

In Section 3.4 we showed that realistic bandwidth constraints can also impose an analytic tradeoff upon sensitivity properties of a feedback system. Bandwidth constraints may arise from several sources, including the natural bandwidth of the plant, high frequency modelling error, and sensor noise. Again, if the sensitivity tradeoff imposed by a bandwidth constraint is too severe, it may be necessary to relax the need for the constraint by changing the plant structure, obtaining a better plant model, and/or reducing sensor noise.

CHAPTER 4

COMMENTS ON DESIGN METHODOLOGIES

4.1. Introduction and Motivation

The two preceding chapters have been devoted to the theory of scalar linear time-invariant feedback systems. Beginning with Chapter 5, we shall seek to develop theory relevant to the analysis and design of multivariable feedback systems. Before embarking upon this course, it is first appropriate to examine classical analysis and design techniques to determine those characteristics that made them successful in addressing the class of design problems for which they were developed. Our motivation in doing so is to obtain information potentially useful in attempts to develop alternative scalar design methodologies and/or methodologies for multivariable design. Of particular interest is the fact that multivariable design methodologies, in contrast to classical methods, will most likely focus directly upon the closed, rather than open, loop transfer functions. Clearly, *any* useful design methodology must possess the ability to cope with the design limitations imposed by the feedback structure, by realizability constraints, and by properties of the plant. Hence it is useful to compare the way in which open and closed loop analysis methods cope with these design limitations. We shall undertake such a comparison by studying two limitations imposed upon design of the system sensitivity function; namely, those due to bandwidth constraints and those due to nonminimum phase plant zeros. Briefly, in classical analysis, these constraints, as well as design specifications imposed upon the sensitivity function, are studied by relating them to open loop gain and phase, which are then used as design parameters. When these properties are studied using closed loop analysis methods, the design specifications are stated directly as requirements upon the magnitude of the sensitivity function. The design limitations, on the other hand, manifest themselves as integral relations which constrain the *relative* magnitude of the sensitivity function in different frequency ranges, and these may preclude satisfaction of the design specifications. We shall now elaborate upon these points.

Recall the three characteristics of classical analysis methods discussed in Chapter 2. The first of these was that certain closed loop transfer functions (in particular, the sensitivity function) are important in that they directly express the quality of a feedback design. It is therefore possible to state design specifications explicitly in terms of these transfer functions.

The second important characteristic of classical theory is that the relation between open and closed loop system properties is reasonably well understood. Such understanding is important for at least two reasons. First, classical design methods involve shaping the gain and phase of the open loop transfer function in order to manipulate feedback properties. Clearly, for this strategy to be successful, it is necessary that feedback properties can be approximated from the values of open loop gain and phase. As we saw in Chapter 2, classical feedback theory provided several rules of thumb, based upon the basic equations governing system behavior, which relate open and closed loop properties and are therefore useful in design.

Even if open loop gain and phase were not used as design parameters, it would still be necessary to understand the relation between open and closed loop system properties. This is because there are several properties of the open loop transfer function which constrain the closed loop system and may, in fact, impose tradeoffs between feedback design goals. Examples of such properties include nonminimum phase zeros, unstable poles, and bandwidth limitations. In particular, when the plant has nonminimum phase zeros and/or unstable poles, these cannot be cancelled with compensation and thus must also be present in the open loop transfer function. Furthermore, as the discussion of (2.3.4-5) showed, there are practical limits upon the extent to which the bandwidth of the open loop system can exceed that of the plant. Hence an understanding of the limitations imposed by these open-loop properties is essential in assessing the relative degree of design difficulty associated with a given plant. If the plant is judged too difficult to control, then it may be necessary to redesign it so that an acceptable feedback control system may be designed.

The preceding discussion has already touched upon a third important characteristic of classical feedback theory — the recognition that various design limitations exist and that these limitations often manifest themselves as tradeoffs between conflicting design goals. Some of these tradeoffs may be quantified directly in terms of the closed loop transfer functions; for example, the conflict at each frequency between disturbance response and noise response. Other tradeoffs, such as that imposed by the limit on the rate of gain rolloff (Section 2.5), are expressed in terms of open loop gain and phase and thus only indirectly in terms of feedback properties. Tradeoffs due to plant properties can also be explored by considering the effect that they have upon the open loop transfer function. For example, as we saw in Chapter 3, the design limitations imposed by right half plane zeros can be analyzed by considering the effect such zeros have upon open loop phase (Figure 3.2.2).

Classical feedback theory provided two types of insight into the limitations inherent in a given design problem. The first of these was the set of mathematical equations, and associated rules of thumb, which were useful in determining *a priori* the degree of design difficulty. For example, we know *a priori* that responses to disturbances and to noise cannot both be small at the same frequency. This knowledge can be useful both in formulating a design specification with compatible objectives as well as in obtaining a reasonable initial choice of a compensator.

A second source of insight into design tradeoffs is the *a posteriori* information obtained in the course of the (iterative) design procedure itself. Indeed, classical methodology may be viewed in part as a means of exploring the set of designs which are achievable given the constraints inherent in the plant and the general properties of linear time invariant systems.

Graphical analysis methods, such as Nyquist and Bode plots, provide considerable insight, not only into whether a given design meets specifications, but also into how the design might be altered to improve some of its properties, if possible. Of course, attempting to satisfy one specification may necessitate that another is violated; insight into such tradeoffs is also provided by classical methods. To illustrate, consider the problem of obtaining a design with satisfactory sensitivity properties when the plant has a nonminimum phase zero (Section 3.2). As we saw, the Nyquist plot (Figure 3.2.2) reveals that use of feedback to decrease sensitivity implies that there necessarily exists a frequency range over which sensitivity is increased. It is clear that increasing the loop gain by a factor $k > 1$ in order to decrease low frequency sensitivity will necessarily cause sensitivity to become even larger at higher frequencies. Of course, one could attempt to circumvent this difficulty by using more complex compensation schemes; for example, using lead or lag compensation to change the gain by different amounts at different frequencies. However, there exist theoretical limitations upon one's ability to alter properties of a design with compensation. For example, the Bode gain-phase relation shows that the gain of a system cannot generally be altered without also changing the phase.

To summarize, classical feedback theory, by providing *a priori* insights into design limitations, can be used to determine an initial choice of a compensation scheme. Once such an initial choice is found, classical iterative design procedures are useful in exploring the neighborhood of this initial choice to attempt a better design. This exploration is guided, both by graphical analysis procedures and by the classical theory of design limitations. The information obtained by these iterative methods is *a posteriori* in the sense that it is obtained after several trial designs have been examined.

Our discussion of classical design and analysis techniques suggests that alternate design methodologies, including those intended for use with multivariable systems, should possess (at least) three features. First, they should provide reliable measures of design quality which may be used to analyze properties of a feedback design and upon which design specifications may be imposed. These measures may include frequency responses of various transfer functions, time responses, or the optimal value of a cost function. Second, the relation between feedback properties and the parameters available for manipulation by the designer should be adequately understood. As we have seen, in classical methodologies the available design parameters are the gain and phase of the open loop transfer function. Alternately, these design parameters may be closed loop pole locations, as in pole placement design techniques, or the weighting functions in optimization methods. Finally, the design theory should provide useful characterizations of the limitations and tradeoffs inherent in a given design problem. One would hope that the theory could provide both *a priori* insights, to be used to obtain an initial design, as well as *a posteriori* insights, to aid in iteratively changing the initial design to achieve a better compromise between conflicting design specifications. In particular, the relation between the design parameters and feedback properties should be sufficiently well understood that the set of possible designs can be explored to find one achieving a satisfactory compromise among conflicting specifications.

In the preceding discussion, we have focussed upon some general features of classical design methodologies which we feel should be incorporated into an alternative design theory, if it is to be successful. It is also important to study the way in which classical methods dealt with specific design limitations, such as bandwidth constraints and nonminimum phase zeros. Clearly, alternative theories must also provide effective means of coping with such limitations. The remainder of this chapter will be devoted to comparing classical design methods, based upon shaping open loop gain and phase, with an alternate "design method" which we shall postulate for the purposes of discussion. This alternate design methodology involves choosing a stable transfer function $S^*(s)$ whose properties satisfy the design goals imposed upon the system sensitivity function. The compensator yielding $S^*(s)$ as the sensitivity function for the plant $P(s)$ can then be found from

$$F^*(s) = P^{-1}(s)[1 - S^{*^{-1}}(s)] \quad . \tag{4.1.1}$$

Motivation for this procedure lies in the fact that the sensitivity function expresses many important properties of a feedback design *directly*, rather than only indirectly, as does the open loop transfer function. Indeed, (4.1.1) has previously been used in the context of a feedback design methodology. An earlier comparison of feedback design via (4.1.1) vs. design by open loop shaping was performed by Horowitz ([Hor63], Section 6.13).

In Section 4.2, we shall consider limitations imposed upon the class of achievable sensitivity functions by the existence of a constraint upon closed (or open) loop bandwidth. We shall show how this limitation may be analyzed using the results of Section 3.4, and then postulate a relation between the resulting tradeoff and that imposed by the Bode gain-phase relation. Section 4.3 will treat a similar problem, using the results of Section 3.2 to discuss the problem of choosing the sensitivity function directly when the plant is nonminimum phase. A summary and conclusions are presented in Section 4.4.

4.2. Sensitivity Design with Bandwidth Constraints

We have seen that the sensitivity function of a feedback system *directly* expresses many important properties of that system, and have thereby been led to consider design specifications of the form (2.3.1-2):

$$|S(j\omega)| \leq M_S(\omega) \, , \quad \forall \; \omega \quad , \tag{4.2.1a}$$

and

$$
\begin{aligned}
|1 - S(j\omega)| &= |T(j\omega)| \\
&\le M_T(\omega), \quad \forall\ \omega \ .
\end{aligned}
\tag{4.2.1b}
$$

Typically, the bound $M_S(\omega)$ is very small at low frequencies to obtain (nominal) sensitivity reduction, while $M_T(\omega)$ is very small at high frequencies. Both bounds are required to be reasonably small, although possibly greater than one, at all frequencies. Note that (4.2.1b) may be viewed as a bandwidth constraint imposed upon the complementary sensitivity function. As we saw in Chapter 2, such a constraint may be needed to provide stability robustness, to reduce the response to noise, and/or to avoid exceeding the natural bandwidth of the plant by an excessive amount.

Suppose, then, that one attempts to select a feedback compensator by finding a stable transfer function $S^*(s)$ satisfying (4.2.1). Assuming that the plant is minimum phase and stable (so that we need not worry about unstable pole-zero cancellations) the compensator can be found readily by solving (4.1.1). However, one must first be certain that a function S^* satisfying (4.2.1) exists. Clearly, it is *necessary* that the bounds (4.2.1) are compatible with the identity

$$
S(s) + T(s) \equiv 1 \ . \tag{4.2.2}
$$

Hence $M_S(\omega)$ and $M_T(\omega)$ cannot both be very small at the same frequency.

Unfortunately, requiring that (4.2.1) be compatible with (4.2.2) is not *sufficient* to guarantee that a stable transfer function satisfying both these bounds exists. As we shall see, in addition to the tradeoff imposed by (4.2.2) between the values of $|S(j\omega)|$ and $|T(j\omega)|$ at *each* frequency, there also exist tradeoffs between the values of $|S(j\omega)|$ and $|T(j\omega)|$ in *different* frequency ranges. We shall now demonstrate this tradeoff using the results of Section 3.5. Next, we shall show how the tradeoff may also be analyzed using a result which was known classically; namely, the Bode gain-phase relation.

In order to achieve sensitivity reduction at low frequencies and stability robustness against high frequency modelling errors, it is realistic to suppose that the bounds (4.2.1) require

$$|S(j\omega)| \leq \alpha < 1 \ , \ \forall \ \omega \leq \omega_1 \tag{4.2.3a}$$

and

$$|T(j\omega)| \leq \varepsilon \left[\frac{\omega_c}{\omega} \right]^{1+k} , \ \forall \ \omega \geq \omega_c > \omega_1 \ , \tag{4.2.3b}$$

where $\varepsilon < 1/2$ and $k > 0$. As we commented at the close of Section 3.4, Corollary 3.4.3 holds with the bound on $|L(j\omega)|$ replaced by one of the form (4.2.3b). Hence there necessarily exists a peak in sensitivity at some frequency $\omega \in (\omega_1, \omega_c)$. Assuming, for simplicity, that $L(s)$ is stable, this peak is bounded below by

$$\sup_{\omega \in (\omega_1, \omega_c)} \log|S(j\omega)| \geq \frac{1}{(\omega_c - \omega_1)} \left\{ \omega_1 \cdot \log(1/\alpha) - \frac{3\varepsilon\omega_c}{2k} \right\} \tag{4.2.4}$$

The existence of this peak may well mean that the specification (4.2.1) cannot be satisfied in the frequency range (ω_1, ω_c). If the peak is too large, then it may be necessary to relax one of the specifications (4.2.3) and thus to perform a tradeoff between the value of $|S(j\omega)|$ at low frequencies and that of $|T(j\omega)|$ at high frequencies. Hence choosing a sensitivity function $S^*(s)$ to satisfy both (4.2.1a) and (4.2.1b) may not only be difficult but impossible. Clearly, the problem of deriving an achievable design specification of the form (4.2.1) is much more difficult than merely insuring that the specification is compatible with the algebraic tradeoff imposed by (4.2.2). Some insight into this problem may be obtained from the right hand side of (4.2.4); however, additional research is needed to understand this tradeoff and to obtain further quantitative estimates of its severity. Additional motivation for attacking such problems is provided by the fact that several authors have proposed and studied optimization problems for which the cost function is a weighted combination of $S(j\omega)$ and $T(j\omega)$ (e.g.,[SLH81, Kwa83, Kwa85, VeJ84, Ste85, OYF86]). Bounds such as (4.2.4) may provide insight into the properties of the solutions to such problems.

As remarked above, the tradeoff just described has long been recognized, albeit somewhat imprecisely and indirectly, from a classical perspective. Indeed, consider the Bode gain-phase relation discussed in Section 2.5. Recall that the problem of satisfying specifications such as (4.2.1) and (4.2.3) could be translated into a problem of shaping the open loop transfer function so that loop gain is sufficiently large at low frequencies, to achieve sensitivity reduction, and sufficiently small at high frequencies, to achieve robustness against high frequency uncertainty. Indeed, the specification (4.2.3) may be used to show that the open loop gain must satisfy the bounds

$$|L(j\omega)| \geq (1/\alpha) - 1 \quad , \quad \forall\ \omega \leq \omega_1 \quad , \tag{4.2.5a}$$

and

$$|L(j\omega)| \leq \varepsilon \left[\omega_c/\omega \right]^{1+k} \Big/ \left[1 - \varepsilon \left[\omega_c/\omega \right]^{1+k} \right] \quad , \quad \forall\ \omega \geq \omega_c \quad . \tag{4.2.5b}$$

Note that the constraint (4.2.3b) upon the bandwidth of the complementary sensitivity function translates readily into a constraint upon open loop bandwidth. In particular, the gain must satisfy $|L(j\omega_1)| \geq (1/\alpha) - 1$ and $|L(j\omega_c)| \leq \varepsilon/(1-\varepsilon)$. If $\alpha\varepsilon \ll 1$, then the loop gain must be decreased by a large amount over the interval (ω_1, ω_c). However, as we discussed in Section 2.5, the Bode gain-phase relation implies that there exists a limit upon the rate at which gain can be decreased in the vicinity of crossover. Essentially, this relation shows that a large rate of gain decrease is necessarily accompanied by a large negative phase lag, which in turn can cause closed loop instability, or at least a very small stability margin. Hence gain cannot be decreased arbitrarily rapidly near crossover, and therefore one of the specifications (4.2.5a-b) may have to be relaxed. Since these specifications were imposed so that the constraints (4.2.3a-b) are satisfied, it follows that there exists a tradeoff between the values of $|S(j\omega)|$ and $|T(j\omega)|$ in different frequency ranges. Hence the tradeoff under discussion can indeed be analyzed, if only indirectly, using classical tools.

As we noted in Section 2.5, analysis of the gain-phase relation becomes more problematic when piecewise constant gain functions are considered. Furthermore, the distinction between systems which are conditionally and unconditionally stable needs further clarification. Indeed, since open loop gain and phase are only indirectly related to feedback properties, one might suspect that a more useful and precise formulation of the tradeoff imposed by the gain-phase relation could be obtained directly in terms of the closed loop transfer functions. Although a complete answer to this conjecture is as yet unavailable, it appears that the Bode sensitivity integral and bounds such as (4.2.4) may provide useful information.

To summarize, when realistic high-frequency bandwidth constraints are imposed upon the complementary sensitivity function and/or the open loop transfer function, it follows that a tradeoff is imposed upon sensitivity properties at lower frequencies. We have postulated that this tradeoff is, at least in part, identical to the classical tradeoff upon the rate of gain rolloff near crossover frequency described by the Bode gain-phase relation.

4.3. Sensitivity Design with Nonminimum Phase Zeros

In the preceding section we assumed, for simplicity, that the plant was minimum phase and stable. Hence, given a desired stable sensitivity function $S^*(s)$, we could always find a feedback compensator $F^*(s)$ yielding an internally stable design. In the present section, we shall remove this assumption and devote our attention to design when the plant has one or more nonminimum phase zeros. For simplicity of exposition, we shall continue to assume that the plant is stable.

Suppose first that the plant has one (real) zero, $z = x$, in the open right half plane. As we saw in Section 3.2, this implies that the sensitivity function must satisfy the constraint

$$S(x) = 1 \quad . \tag{4.3.1}$$

Consider the design specification

$$|S(j\omega)| \le M_S(\omega) \ , \ \forall \ \omega \quad , \tag{4.3.2a}$$

where

$$M_S(\omega) \le \alpha < 1 \ , \ \forall \ \omega \le \omega_l \tag{4.3.2b}$$

in order to provide low-frequency sensitivity reduction. We shall also require $M_S(\omega)$ to be reasonably small (but possibly greater than one) at higher frequencies.

As we saw in Section 3.2, the algebraic constraint (4.3.1) has an equivalent statement

$$\int_0^\infty \log|S(j\omega)| \ W(x,\omega) d\omega = 0 \quad . \tag{4.3.3}$$

Together, (4.3.2b) and (4.3.3) may be used, as in Corollary 3.2.2, to show that there necessarily exists a peak in sensitivity which may violate the bound (4.3.2a) at frequencies $\omega > \omega_l$. Hence it is always necessary to ascertain whether the constraint (4.3.1) and the design goals (4.3.2) are compatible conditions.

Consider, as in the previous section, an approach to design whereby we choose a sensitivity function $S^*(s)$ to satisfy (4.3.2) and then calculate the associated compensator from

$$F^*(s) = P^{-1}\left[1 - S^{*^{-1}}(s)\right] \quad . \tag{4.3.4}$$

Clearly, unless $S^*(x) = 1$, $F^*(s)$ will have an unstable pole at $s = x$ and the system will not be internally stable. Let us therefore postulate, for sake of discussion, an iterative "design procedure" whereby we choose candidate sensitivity functions $S_i(s)$ satisfying the design goal

(4.3.2) and iterate until the constraint (4.3.1) is also satisfied. To do this it may, of course, be necessary to relax the bound (4.3.2). To illustrate, let $S_1(s)$ be a candidate sensitivity function. Then, from Poisson's integral theorem, it follows that

$$\pi \log |S_1(x)| = \int_0^\infty \log |S_1(j\omega)| W(x,\omega) d\omega \quad . \tag{4.3.5}$$

This equation is significant since a compensator yielding an internally stable feedback design exists only if $S_1(x) = 1$. Now, suppose that this latter condition is violated; in particular, suppose that $S_1(x) < 1$. (Since good sensitivity is achieved by requiring $|S_1(j\omega)|$ to be sufficiently small, this is surely the most likely case, at least in an initial design attempt.) Hence it will be necessary to choose a new candidate sensitivity function $S_2(s)$ whose gain is *larger* than that of $S_1(s)$ over some frequency range. The weighting function $W(x,\omega)$ yields at least some insight into how this might be done. For example, choosing a function $S_2(s)$ whose gain is larger than that of $S_1(s)$ only at lightly weighted frequencies will not generally be effective; the gain of the sensitivity function must be increased at frequencies for which the weighting is significantly large. If it turns out that the weighting is large only within the frequency range over which sensitivity is desired to be small, then a nontrivial relaxation of the design specification will have to be made.

Even with the insight available from (4.3.5), it will be a difficult task to modify an existing $S_i(s)$ to obtain a function more nearly satisfying the constraint (4.3.1) without unnecessarily sacrificing the design goal (4.3.2). As we shall now see, the fact that $P(s)$ was assumed to have only *one* nonminimum phase zero greatly facilitates this procedure. Given a candidate sensitivity function $S_i(s)$ which fails to satisfy (4.3.1), one can easily obtain a new function which does satisfy (4.3.1) simply by scaling:

$$S_{i+1}(s) \triangleq S_i(s)/S_i(x) \quad . \tag{4.3.6}$$

Of course, if $S_i(x) \ll 1$, then $|S_{i+1}(j\omega)|$ will be much greater than $|S_i(j\omega)|$ at all frequencies. Suppose, however, that we are able to find a function $S_i(s)$ which satisfies the design goal (4.3.2) reasonably well and for which $S_i(x) \approx 1$. (Of course, finding such a function may be nontrivial, even with the insights furnished by (4.3.5).) Then $S_{i+1}(s)$ will not only yield an internally stable design, but will approximately satisfy (4.3.2) as well.

We have just shown that if the plant has only one nonminimum phase zero, then simple scaling suffices to modify a candidate sensitivity function to obtain a new function satisfying the integral constraint (4.3.3); simply choose $S_{i+1}(s) = kS_i(s)$, where $k = 1/S_i(x)$. It is easy to show that, if the plant has $N_z > 1$ nonminimum phase zeros, then scaling a candidate sensitivity function will not generally suffice to allow all the N_z integral constraints to be satisfied. Indeed, even a complex conjugate pair of zeros yields two integral constraints upon $|S(j\omega)|$; namely, equations (3.2.16) and (3.2.19). Suppose, however, that we are willing to use unstable poles in the compensator.[7] It turns out that, if we allow the compensator to possess N_z-1 appropriately placed unstable poles, then the integral constraints can all be satisfied. To illustrate, suppose that $P(s)$ has two real nonminimum phase zeros, z_1 and z_2, and let $S_i(s)$ be a candidate sensitivity function with $S_i(z_1) \neq 1$ and/or $S_i(z_2) \neq 1$. Define a new candidate function $S_{i+1}(s) = kS_i(s)B_F(s)$, where $B_F(s) = \frac{p-s}{p+s}$ is the Blaschke product containing the nonminimum phase zero of $S_{i+1}(s)$ located at the unstable compensator pole.[8] Then, at each zero

---

[7]To simplify the discussion, we continue to assume that the plant is stable.

[8]Since $P(s)$ is assumed stable, this is the only such zero of $S_{i+1}(s)$.

z_l, $l = 1,2$, the function $S_{i+1}(s)$ must satisfy the Poisson integral:

$$\frac{1}{\pi}\int_0^{\infty} \log |S_{i+1}(j\omega)| W(z_l,\omega) d\omega = \log \left| \frac{p+z_l}{p-z_l} \right| - \log |k| + \log |S_i(z_l)| \quad . \tag{4.3.7}$$

It is straightforward to show that p and k may be chosen so that the right hand side of (4.3.7) equals zero for both z_1 and z_2. Hence $S_{i+1}(z_1) = S_{i+1}(z_2) = 1$, and the compensator calculated from (4.3.4) with $S^*(s) = S_{i+1}(s)$ will yield an internally stable design. Of course, as in the case of one zero, the value of $|S_{i+1}(j\omega)|$ may be unacceptably large. A further potential difficulty is due to the use of an unstable compensator to stabilize a stable plant. This results in a system which is (unnecessarily) open loop unstable and therefore only conditionally stable. Decreases in gain due to parameter variations or the nonlinear effects of even temporary saturation may result in closed loop instability. Furthermore, unstable poles impose additional design tradeoffs upon the complementary sensitivity function (Section 3.3) and actually worsen the tradeoffs imposed by nonminimum phase zeros (Section 3.2) and bandwidth constraints (Section 3.4).

When the plant has more than two nonminimum phase zeros, then one can show that the strategy discussed above can be extended using a general interpolation procedure such as that discussed in ([ZaF83], Section VII). It is worth remarking that this problem results in a design which is identical to that resulting from H^∞-sensitivity minimization [ZaF83] for an appropriate choice of weighting function. See [FrL86c] for further discussion.

There exists an alternative to using unstable compensators which is interesting at least mathematically. Recall our remark, following Theorem 3.2.1, that zeros of the sensitivity function on the $j\omega$-axis (at corresponding poles of the compensator) do not effect the value of the Poisson integral. If $S(s)$ has an essential singularity on the $j\omega$-axis, however, then this does contribute to the value of the integral. Hence using a compensator which has no poles in the open right half plane, but has an appropriate number of essential singularities on the $j\omega$-axis, provides an alternate strategy to satisfy the Poisson integral constraints. Indeed, posing H^∞ sensitivity optimization problems with the compensator constrained to have no unstable poles results in an optimal compensator which does have essential singularities on the $j\omega$-axis[Hel85]. Implementation of compensators with such wild discontinuities (or even approximations to such compensators) may not be practically feasible.

We close this section by commenting briefly upon the way that classical design methods deal with the difficulties discussed above. First, since the design variable in open loop-shaping techniques is the compensator itself, the whole problem of avoiding unstable pole-zero cancellation is avoided easily. Of course, closed loop stability is not guaranteed, and the compensator must still be chosen to obtain a satisfactory compromise between sensitivity properties at different frequencies. As we saw in Section 3.2, the relation between open loop gain and phase and sensitivity properties is sufficiently well understood that classical techniques provide an effective way to perform this tradeoff.

4.4. Summary and Conclusion

The purpose of this chapter has been to determine some characteristics of classical feedback theory which made it successful in dealing with a large class of scalar design problems. Our motivation is to find guidelines for the development of alternative scalar design methodologies as well as methodologies for multivariable design. We first considered this question on a very general level, with the results summarized in Section 4.1. Next, we turned our attention to two specific design problems; namely, sensitivity design when the system must satisfy closed (or open) loop bandwidth constraints and when the plant is nonminimum phase. In each case, the limitations and tradeoffs could be analyzed, albeit indirectly and imprecisely, using classical Nyquist and Bode techniques. Specifically, the fact that high frequency bandwidth constraints impose a tradeoff upon low-frequency sensitivity reduction may be observed from the Bode gain-phase relation. In addition, as we saw in Section 3.2, standard Nyquist analysis reveals qualitatively the limitations upon sensitivity imposed by the extra phase lag due to nonminimum phase zeros. Although the classical treatment of each of these tradeoffs involves open loop gain and phase in an essential way, we have shown that the results of Chapter 3 may be used to study these design problems using the closed loop transfer functions directly.

An obvious limitation of the classical design methodologies for scalar systems is that they are not readily applicable to systems with more than one input and/or output. One consequence of the attempts to develop analysis and design methods for multivariable systems has been that the classical techniques themselves have been reevaluated from a more rigorous standpoint. In particular, renewed emphasis has been placed upon the primary role played by the closed, rather than the open, loop transfer functions. For example, it has been shown that the classical notions of gain and phase margin do not necessarily provide adequate characterizations of stability robustness (e.g.,[LCL81, AsH84]). Indeed, these margins are only approximations to the notion of closest distance to the critical point, which in turn is just the reciprocal of the system sensitivity function (Figure 3.2.1). This fact, along with other considerations, suggests that design methodologies for multivariable systems will probably focus primary attention upon manipulating closed loop properties *directly*, rather than indirectly as did classical loop-shaping design methods. Since the closed loop transfer functions *do* directly express the quality of a design, it is difficult to argue with this line of reasoning. Hence, we feel that efforts should be made to discover how these tradeoffs are displayed using analysis techniques based upon the closed loop transfer functions. In particular, this project holds potential for clarifying limitations that were only imprecisely stated in terms of the open loop transfer functions. Toward this end, the use of Poisson's integral to study the sensitivity tradeoff due to nonminimum phase zeros appears to be a useful step. Furthermore, the fact that rate of gain rolloff problems can be studied using the Bode sensitivity integral and bounds such as (4.2.6) promises to give this class of design tradeoffs a more explicit formulation. One particularly intriguing aspect of this problem arises from the somewhat uncertain role played by phase in multivariable systems (see Chapter 6). Suppose that those classical results which appear to involve phase in an explicit way (such as the tradeoff imposed by the Bode gain-phase relation) can be given equivalent statements not involving phase directly. Then perhaps these equivalent statements could be generalized to multivariable systems without first having to develop a useful notion of multivariable phase.

In addition to the need for analysis methods, there is also a need to develop closed loop design methods which are effective in manipulating the various design tradeoffs. Obviously, we cannot naively choose a sensitivity function as we postulated, for the sake of discussion, in this chapter. One potential way to resolve this difficulty is to use interpolation theory to parameterize the set of internally stabilizing compensators (see [Vid85] and references therein). Once this parameterization has been carried out, there remains the task of choosing one of these internally stable designs which also has satisfactory feedback properties. Presumably, the free parameter (a stable transfer function) which appears in this parameterization can be manipulated to obtain a reasonable design; methods of obtaining insight into the relations between this parameter and feedback properties should be developed to aid in this procedure (see[CaD82], pp. 224-225, for an example). One would hope that, in the scalar case, rules of thumb could be developed which are as insightful as those relating feedback properties to open loop gain and phase when these variables are used as design parameters. In any event, it will greatly facilitate the design process if the free parameters prove to be effective in exploring the set of feedback designs possible given the constraints imposed by the problem. Of course, one way to choose the free parameter is by solving an optimization problem. Clearly, the success of this procedure will depend critically upon one's ability to formulate an optimization problem which incorporates all the various goals present in a practical design problem. Omission of one of these goals, even one of minor importance, may mean that it is violated excessively by the resulting design. Alternately, one could view optimization based design methods merely as tools for iterative design. In this regard, much work has been done in relating the free parameters in the LQG/LTR design methodology to feedback system properties; see the tutorial [StA84] for a thorough discussion. Yet another useful tool for gaining insight into design tradeoffs is furnished by the methods of O'Young and Francis ([OYF85, OYF86]).

It is beyond the scope of this chapter and this monograph to survey the large and ever increasing number of design methodologies proposed in the literature. We have merely sought to raise some questions and to point out some issues which should be addressed by any methodology, if it is to be successful. To summarize, in addition to mathematically rigorous design theories and efficient computational procedures, we also need analysis methods to obtain more qualitative insights into the difficulties posed by a given design problem. After these have been analyzed, effective methods are also needed to find a design, if one exists, which achieves a reasonable compromise among various design goals. It is a fact that classical design methods, although limited in scope, did prove to be successful in many applications. This observation suggests that classical methods be studied to ascertain the reason for their effectiveness and to provide guidelines for the development of more generally applicable design methodologies.

CHAPTER 5

MULTIVARIABLE SYSTEMS: SUMMARY OF EXISTING RESULTS AND MOTIVATION FOR FURTHER WORK

5.1. Introduction

In Chapters 2-4 we examined some important concepts from the classical theory of scalar linear feedback systems. Among these were that closed loop transfer functions directly express the quality of a feedback design, that the relation between open and closed loop system properties is readily understood, and that design tradeoffs must be made both between system properties at a fixed frequency and between system properties in different frequency ranges. We feel that these concepts are sufficiently useful and interesting that they merit generalization to multivariable systems.

Indeed, work on generalizing classical design and analysis methods to multivariable systems has been in progress for many years (e.g., [Ros74] and[PoM79]). Beginning in the late 1970's this area received renewed interest. The results of the first few years of this renewed activity are described in the tutorial summaries [DoS81, SLH81, LSA81, PEM81, CFL81]. Of these, [SLH81] and, especially, [DoS81] contain results which are the most relevant to this monograph. Of course, many approaches to multivariable system theory exist and we cannot hope to summarize all these. A number of important papers related to frequency response methods are found in the collection [Mac79]. Matrix factorization methods have received much attention; see the monographs [CaD82] and [Vid85]. Much work in recent years has stemmed from the H^∞-optimization approach to system design initiated in [Zam81] and surveyed in [FrD85]. A number of design limitations are studied in [KhT85].

Our first task in this chapter is to summarize some of these early generalizations of scalar concepts to multivariable systems. We shall then describe limitations on the applicability of these generalizations; essentially, they provide useful information only in special cases for which multivariable systems behave sufficiently like scalar systems. It is becoming increasingly clear, however, that multivariable systems can possess many properties having no direct scalar analogues. Those properties of interest in this monograph arise because multivariable systems and signals exhibit *directionality* properties.

Loosely speaking, directionality refers to any property of a matrix transfer function or vector signal which exhibits a *spatial* dependency in addition to the *frequency* dependency shared with scalar transfer functions and signals. For example, a matrix transfer function evaluated at a pole only "blows up" in directions determined by the residue matrix at the pole. Likewise, multivariable zeros only block transmission of signals lying in the null-space of the transfer function evaluated at the zero; generally, this null-space is only a proper subspace of the input space.

Typically, disturbance and noise signals do not appear at uniform levels in all loops of a multivariable system, but affect some loops more than others. Hence, we shall frequently speak of the *direction* of a vector signal; by this, we mean the one-dimensional subspace of $\mathbb{C}^n$ in which the signal lies. When convenient, we shall refer to the *loops* of the system, and identify these with the standard basis directions.

Multivariable systems display directionality properties in that they respond differently to signals lying in different directions. It is not hard to find examples of plants whose transfer function matrices have both large and small gain in different directions at the same frequency. As we shall see in Section 5.3, this behavior can be made precise using the singular value decomposition of the transfer function matrix. In particular, it is important to recognize that uncertainty in multivariable systems displays directionality properties. For example, one loop of a system may contain substantially more uncertainty due to unmodelled dynamics or parameter variations than do other loops.

In the remainder of this chapter, we shall discuss why straightforward generalizations of classical concepts fail to adequately cope with phenomena related to directionality, and shall describe some classes of multivariable design specifications in which directionality plays an important role. We thus provide motivation for the work to be pursued in the remainder of the monograph.

5.2. Quantification of Multivariable Feedback Performance

Consider the linear time-invariant feedback system shown in Figure 5.2.1. This configuration is the same as that shown in Figure 2.2.1, with the exception that signals are vector-valued and the plant and compensator transfer functions are matrices.[9] We shall assume that $P(s) \in \mathbb{C}^{n \times m}$, $F(s) \in \mathbb{C}^{m \times n}$, $n(s) \in \mathbb{C}^{n}$, $d(s) \in \mathbb{C}^{n}$, $y(s) \in \mathbb{C}^{n}$, $r(s) \in \mathbb{C}^{m}$, and $u(s) \in \mathbb{C}^{m}$.

Since matrix multiplication is in general noncommutative, breaking the loop at the plant input (point (x) in Figure 5.2.1) generally yields a different open loop transfer function than is obtained by breaking the loop at the plant output (point (xx)). Hence, in the multivariable case, two sets of transfer functions correspond to the scalar functions (2.2.1-3). Let the open loop transfer function, sensitivity function, and complementary sensitivity function at the plant

---

[9]We assume that the units in which different signals are measured have been chosen to be physically reasonable. In general, the problem of scaling in a multivariable system can be quite problematic.

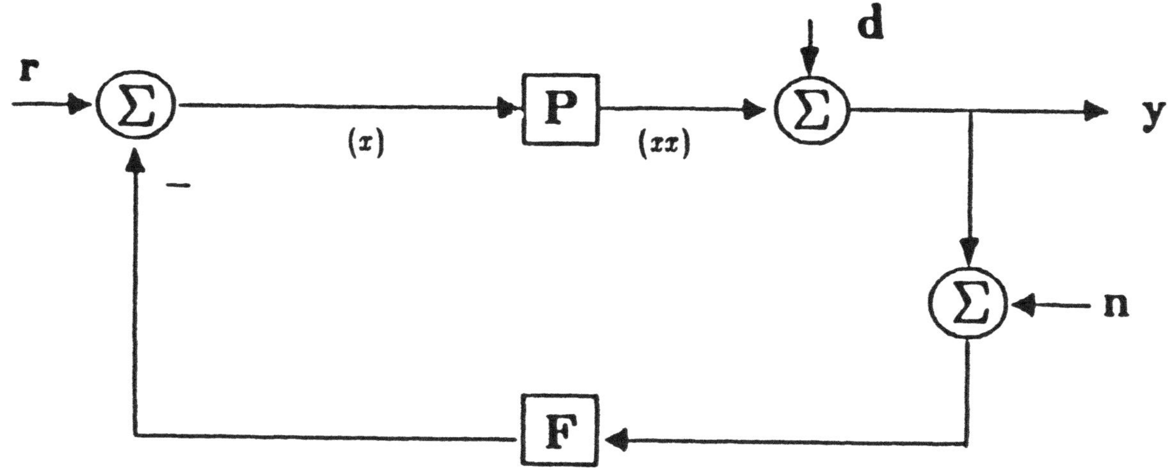

Figure 5.2.1: Multivariable Feedback System

output be denoted

$$L_O(s) \triangleq P(s)F(s) \tag{5.2.1}$$

$$S_O(s) \triangleq [I+L_O(s)]^{-1} \tag{5.2.2}$$

$$T_O(s) \triangleq L_O(s)[I+L_O(s)]^{-1} \tag{5.2.3}$$

and denote the corresponding transfer functions at the plant input by

$$L_I(s) \triangleq F(s)P(s) \tag{5.2.4}$$

$$S_I(s) \triangleq [I+L_I(s)]^{-1} \tag{5.2.5}$$

$$T_I(s) \triangleq L_I(s)[I+L_I(s)]^{-1} \tag{5.2.6}$$

The output of the system can again be decomposed into the sum of three terms, each of which is governed by one of the transfer functions defined at the plant output:

$$y(s) = y_r(s)+y_d(s)+y_n(s) \tag{5.2.7}$$

where

$$y_r(s) = S_O(s)P(s)r(s) \tag{5.2.8}$$

$$y_d(s) = S_O(s)d(s) \tag{5.2.9}$$

and

$$y_n(s) = -T_O(s)n(s) \quad . \tag{5.2.10}$$

The response of the plant input to commands, noise, and disturbances is given by

$$u(s) = u_r(s) + u_d(s) + u_n(s) \quad , \tag{5.2.11}$$

where

$$u_r(s) = S_I(s)r(s) \tag{5.2.12}$$

$$u_d(s) = -F(s)S_O(s)d(s) \tag{5.2.13}$$

and

$$u_n(s) = -F(s)S_O(s)n(s) \quad . \tag{5.2.14}$$

Note that the response of the plant input to commands is governed by the input sensitivity function. Since, as we shall see shortly, the values of S_I and S_O may differ significantly, it is necessary to evaluate both of these functions when assessing the quality of a feedback design.

As in the scalar case, if no unstable hidden modes exist, then closed loop stability can be inferred from stability of either the output or the input sensitivity function. Closed right half plane poles and zeros of $P(s)$ and $F(s)$ must appear with at least the same multiplicity in $L_O(s)$ and $L_I(s)$. The directionality properties of these poles and zeros must also be reflected in the open loop transfer functions. For example, if the plant has a zero at $s = z$, then not only must $L_O(s)$ have a zero at z, but $L_O(z)$ must also have a left null-space containing the left null-space of $P(z)$.

If the feedback system is stable, then the steady-state response of the system output to a disturbance $d(t) = d_0 e^{j\omega t}$, where $d_0 \in \mathbb{C}^n$ is constant, is governed by the sensitivity function: $y_d(t) \rightarrow S_O(j\omega)d_0 e^{j\omega t}$. Hence the response to such a disturbance can be made small by requiring that[10] $\| S_O(j\omega)d_0 \| \ll 1$. Clearly, the size of the response to a disturbance will

[10]In this monograph, the standard Euclidean vector norm and induced matrix norm will be used exclusively.

depend upon the direction in which it lies. Recall the definition of the largest singular value of a matrix (Appendix A):

$$\overline{\sigma}[S_O(j\omega)] = \max_{\substack{d_0 \in \mathbb{C}^n \\ ||d_0||=1}} ||S_O(j\omega)d_0|| \tag{5.2.15}$$

It follows that the response of the system to a disturbance $d_0 e^{j\omega t}$ can be made small *regardless* of the direction of the disturbance by requiring $\overline{\sigma}[S(j\omega)] << 1$. If this condition holds, then we say that disturbance response is small in *all* directions at frequency ω. In the event that $\overline{\sigma}[S_O(j\omega)] < 1$ then, as in the scalar case, use of feedback is said to reduce the response to disturbances relative to that of an open loop system. The effect of feedback upon the noise response properties of the system may be quantified using the complementary sensitivity function. Noise response is small in all directions at frequency ω provided that $\overline{\sigma}[T_O(j\omega)] \ll 1$.

The output sensitivity function is also useful in assessing the benefits of feedback in reducing system response to plant parameter variations. Remarks analogous to those following (2.2.12-13) hold in the multivariable case, with absolute values replaced by singular values, and thus will not be dwelt upon here.

Each of the sensitivity and complementary sensitivity functions (5.2.2,3,5,6) characterizes stability robustness of the feedback system against a class of plant uncertainty (e.g., see [DWS82], Table 1.). As in the scalar case, uncertainty at high frequencies due to unmodelled dynamics is frequently described by a multiplicative perturbation. Unlike the scalar case, however, one must distinguish between multiplicative uncertainty at the plant input

$$P'(s) \triangleq P(s)(I+\Delta_I(s)) \tag{5.2.16}$$

and multiplicative uncertainty at the plant output

$$P'(s) \triangleq (I+\Delta_O(s))P(s) \quad . \qquad (5.2.17)$$

For example, uncertainty due to unmodelled actuator dynamics would be more appropriately described by (5.2.16) while uncertainty associated with the sensors would be described by (5.2.17). In each case, the perturbation $\Delta_i(s)$ is assumed to lie in a set of stable transfer functions satisfying a frequency-dependent magnitude bound:

$$\mathbf{D}_i(s) \triangleq \{\Delta_i(s) \text{ is stable and } \bar{\sigma}[\Delta_i(j\omega)] \le M_i(\omega) \ , \ \forall \ \omega\} \qquad (5.2.18)$$

If the only available information about system uncertainty is that it is multiplicative, occurs either before *or* after the plant, and satisfies (5.2.18), then the perturbation $\Delta_i(s)$ is termed *unstructured* uncertainty. Necessary and sufficient conditions for the feedback system to be stable for all plants described by multiplicative *input* uncertainty (5.2.16) are that the system is nominally stable and that the complementary sensitivity function at the plant input satisfy the bound

$$\bar{\sigma}[T_I(j\omega)] < 1/M_I(\omega) \ , \ \forall \ \omega \quad . \qquad (5.2.19)$$

Suppose, on the other hand, that the uncertainty is unstructured but occurs at the plant output (5.2.17). Then the system will be stable for all plants in this class if and only if the system is nominally stable and the complementary sensitivity function at the plant output satisfies the bound

$$\bar{\sigma}[T_O(j\omega)] < 1/M_O(\omega) \ , \ \forall \ \omega \quad . \qquad (5.2.20)$$

As in the scalar case, uncertainty typically becomes large at high frequencies, thus forcing the complementary sensitivity functions to be small at high frequencies to insure that the system is robustly stable.

The fact that robustness against uncertainty in different loop-breaking locations is governed by different transfer functions is responsible for several phenomena unique to multivariable systems. One of these is that robustness against uncertainty at one loop location can be quite good while, at the same frequency, the stability margin against uncertainty in the other location can be extremely small. It turns out that this can occur *only* when both plant and compensator transfer functions have a high condition number ([Ste84, Ste85]). To show this, assume for simplicity that the plant is square and invertible. It is easy to show that

$$T_I(s) = P^{-1}(s) T_O(s) P(s) \quad . \tag{5.2.21}$$

Recalling that the plant condition number is defined by $\kappa[P(s)] = \bar{\sigma}[P(s)]/\underline{\sigma}[P(s)]$ and using well-known properties of the singular value decomposition (Appendix A), it follows that

$$\bar{\sigma}[T_O(j\omega)]/\kappa[P(j\omega)] \le \bar{\sigma}[T_I(j\omega)] \le \bar{\sigma}[T_O(j\omega)]\kappa[P(j\omega)]$$

and (5.2.22)

$$\bar{\sigma}[T_I(j\omega)]/\kappa[P(j\omega)] \le \bar{\sigma}[T_O(j\omega)] \le \bar{\sigma}[T_I(j\omega)]\kappa[P(j\omega)] \quad .$$

These bounds show that, if $\kappa[P(j\omega)] \approx 1$, then the two complementary sensitivity functions have approximately the same size. Hence the associated stability margins are also about equal. Bounds similar to (5.2.22), with $\kappa[P(j\omega)]$ replaced by $\kappa[F(j\omega)]$, show that use of a well-conditioned compensator also guarantees that stability margins will be approximately the same at both loop-breaking points. If both plant and compensator have high condition numbers then, since one can easily construct examples for which these bounds are *tight*, it follows that knowing the stability margin at one point tells us nothing useful about the stability margin at the other point. Similar remarks apply to the two sensitivity functions, as may be seen by using the identities $S_i + T_i = I$ in conjunction with (5.2.22).

A more serious difficulty, also unique to multivariable systems, occurs when plant uncertainty is present simultaneously at both loop-breaking points (as it always is in practice). Suppose that both $\bar{\sigma}[T_O(j\omega)]$ and $\bar{\sigma}[T_I(j\omega)]$ are sufficiently small that the system is robustly stable against perturbations occurring at the plant output *or* the plant input. It turns out that there is still no guarantee that the system cannot be destabilized by small perturbations occurring at both these points simultaneously. An example illustrating this problem is discussed in[FLC82].

Fortunately, Doyle ([Doy82, DWS82]) has developed an analysis tool, the *structured singular value*, which is useful in providing necessary and sufficient conditions for stability robustness in the presence of simultaneous uncertainty.[11] Applied to the simultaneous uncertainty problem, the structured singular value essentially allows one to analyze the effect of uncertainty at one loop-breaking point upon the stability margin against uncertainty at the other point. It is interesting to note that this methodology also requires analysis of a closed-loop transfer function. The transfer function needed is obtained by rearranging the block diagram of the system to reveal how perturbations in different loop locations can couple to produce instability. It may be shown ([Ste84, Ste85]) that the stability margins against individual perturbations can yield misleading estimates of robustness against simultaneous perturbations only if the plant and/or compensator are ill-conditioned.

Our motive in mentioning the problems associated with an ill-conditioned plant is merely to emphasize the potential difficulties associated with design and analysis of multivariable feedback systems. Further discussion of these problems is beyond the scope of this monograph. Some insight into the role that the plant condition number plays in multivariable design can be

---

[11] This analysis is also useful for a much broader class of problems than those we discuss here.

obtained using the results of Chapter 7. The interested reader is referred to [FrL85d] and [FrL86a] for details.

5.3. Generalization of Scalar Concepts to Multivariable Systems

Under certain conditions, multivariable systems behave sufficiently like scalar systems that classical results extend easily. In this section we shall describe what these conditions are, and then discuss the relevant generalizations. Most of these results are well-known and may be found in various references; e.g., [DoS81, SLH81].

We saw at the close of the preceding section that, unlike scalar systems, multivariable systems can possess substantially different stability margins against uncertainty occurring at different loop-breaking points. To avoid worrying about this problem, we shall assume that the plant to be controlled has a well-conditioned transfer function matrix. This assumption will also insure that the system is not extremely sensitive to the effects of simultaneous uncertainty. (Alternatively, we could assume that uncertainty is present at only *one* loop-breaking location. If this approach is taken, then it is also necessary to assume that disturbances enter the system at the same loop-breaking point at which uncertainty occurs ([DWS82, Ste85]).) We shall also assume in this section that levels of uncertainty, noise, and disturbances are approximately equal in all loops of the system. This assumption will allow us to describe the uncertainty using an *unstructured* perturbation; i.e., one for which no knowledge is available other than loop location, stability, and a frequency-dependent magnitude bound. The term "unstructured" will also be used to describe a disturbance or noise signal which is arbitrary except for a frequency-dependent bound on its size. Hence, we assume that the disturbance and noise signal satisfy

$$||d(j\omega)|| \le M_d(\omega) \quad \forall\ \omega \tag{5.3.1}$$

and

$$||n(j\omega)|| \le M_n(\omega) \quad \forall \ \omega \tag{5.3.2}$$

For specificity, we shall assume that the unstructured uncertainty is multiplicative and occurs at the plant output (5.2.17) where disturbances enter the loop. Finally, we shall assume that each of the outputs of the system is equally important, in that the response of each to disturbances and noise is desired to be equally small. Hence, we require that

$$||y_d(j\omega)|| \le M_{y_d}(\omega) \tag{5.3.3}$$

and

$$||y_n(j\omega)|| \le M_{y_n}(\omega) \tag{5.3.4}$$

The assumptions that the plant is well-conditioned and that noise, disturbances, and uncertainty are unstructured are sufficiently strong to rule out inherently multivariable system phenomena. Nonetheless, the class of systems which do satisfy these assumptions is important in practice, and it is satisfying to know that scalar results can be generalized when appropriate.

Recall the three concepts arising in scalar feedback theory which we chose to highlight in Section 2.6. In the preceding section, we saw that the first of these could indeed be generalized to multivariable systems. Specifically, the sensitivity and complementary sensitivity functions are useful in evaluating the quality of a feedback design. This remains true even when uncertainty is structured and occurs at more than one loop location. The key difference is that two sets of transfer functions must be evaluated, (5.2.2-3) *and* (5.2.5-6), as well as the transfer function used in structured singular value analysis [DWS82].

Together, the assumptions that uncertainty, noise, and disturbances are unstructured, and that all outputs are equally important, imply that design specifications can be translated into the bounds

$$\overline{\sigma}[S_O(j\omega)] \le M_S(\omega) \quad , \quad \forall\ \omega \qquad (5.3.5a)$$

and

$$\overline{\sigma}[T_O(j\omega)] \le M_T(\omega) \quad , \quad \forall\ \omega\ . \qquad (5.3.5b)$$

At frequencies for which disturbance rejection and nominal sensitivity reduction are desired, the bound on sensitivity should satisfy $M_S(\omega) \ll 1$. Similarly, at frequencies for which design goals include noise rejection and robustness against large levels of multiplicative uncertainty, the bound on complementary sensitivity should be very small. As in the scalar case, there is a conflict between achieving these two goals at the same frequency. The identity

$$S_O(s) + T_O(s) = I \qquad (5.3.6)$$

implies that $\overline{\sigma}[S_O(j\omega)]$ and $\overline{\sigma}[T_O(j\omega)]$ cannot both be small at the same frequency. Hence, at each frequency, there exists an algebraic tradeoff between the feedback properties governed by S_O and those governed by T_O. The lack of directionality properties and structure in the systems we are considering implies that this tradeoff is essentially the same as that in the scalar case. Algebraic tradeoffs when uncertainty, disturbances, and noise have structure will be treated in the next section.

We also saw in Chapter 2 that the relations between open and closed loop properties of a scalar feedback system are fairly well understood. In special cases, the rules of thumb relating open loop gain to feedback properties can be extended to multivariable systems. Before discussing this, we need to make precise what we mean by the gain of a matrix transfer function.

Let $M(j\omega) \in \mathbb{C}^{n \times n}$ be a transfer function evaluated at frequency ω. The level of gain experienced by an input to the system described by this transfer function will generally depend upon the direction in which the input lies. Using the singular value decomposition (Appendix

A), it is possible to define, at each frequency, a canonical set of gains for $M(j\omega)$. Recalling that

$$\overline{\sigma}[M(j\omega)] \triangleq \max_{\|u\|=1} \|M(j\omega)u\| \tag{5.3.7}$$

and

$$\underline{\sigma}[M(j\omega)] \triangleq \min_{\|u\|=1} \|M(j\omega)u\| \quad , \tag{5.3.8}$$

it follows that the largest and smallest possible values of gain are equal to the largest and smallest singular values. If $\underline{\sigma}[M(j\omega)] \gg 1$, then we say that the gain of the system is large in all directions at frequency ω. If $\overline{\sigma}[M(j\omega)] \ll 1$, then the gain is small in all directions at that frequency. Hence, if $\overline{\sigma}[M(j\omega)] \gg 1 \gg \underline{\sigma}[M(j\omega)]$, we say that the system has both large gain and small gain in different directions at the same frequency.

The input direction associated to the ith singular value gain is determined by the right singular subspace U_i and the output direction is determined by the left singular subspace V_i. The fact that gain depends upon direction is seen most easily by writing the singular value decomposition in dyadic form

$$M = \sum_{i=1}^{n} v_i u_i^H \sigma_i \quad . \tag{5.3.9}$$

If we assume, for simplicity of discussion, that the singular values are distinct, then the singular subspaces V_i and U_i are one-dimensional and are spanned by the singular vectors v_i and u_i. An input signal in the direction of the ith right singular subspace is amplified by the gain σ_i and appears in the output direction given by the ith left singular subspace.

When the open loop transfer function has large gains in all directions or small gain in all directions, the following analogues to (2.4.1-2) are useful in assessing the feedback properties of the system ([DoS81, CaD82]):

$$\underline{\sigma}[L_O(j\omega)] \gg 1 \iff \begin{array}{c} \bar{\sigma}[S_O(j\omega)] \ll 1 \\ \text{and} \\ T_O(j\omega) \approx I \end{array} \tag{5.3.10}$$

and

$$\bar{\sigma}[L_O(j\omega)] \ll 1 \iff \begin{array}{c} S_O(j\omega) \approx I \\ \text{and} \\ \bar{\sigma}[T_O(j\omega)] \ll 1 \end{array} \quad . \tag{5.3.11}$$

Generalizations of scalar rules of thumb follow readily from these approximations. At frequencies for which open loop gain is large in all directions, the feedback system has small disturbance response in all directions and all the noise appears directly in the system output. At frequencies for which open loop gain is small in all directions, the noise response of the system is everywhere small but all disturbances appear directly in the output. If neither of these conditions hold, approximation of feedback properties from those of the open loop system is less straightforward. We shall study this problem further in Chapter 7.

Recall the algebraic design tradeoffs discussed in Chapter 2. We have already seen that the tradeoff between disturbance response and noise response is present in multivariable systems. The algebraic tradeoff between sensitivity reduction and the magnitude of the control input is also present. Assuming for the moment that the plant is square and invertible, (5.2.13) implies that

$$\begin{aligned} u_d(s) &= -F(s)S_O(s)d(s) \\ &= -P^{-1}(s)[I - S_O(s)]d(s) \quad . \end{aligned} \tag{5.3.12}$$

Hence, at frequencies for which $\bar{\sigma}[S_O(j\omega)] \ll 1$, we have

$$\|u_d(j\omega)\| \approx \bar{\sigma}[P^{-1}(j\omega)]\,\|d(j\omega)\| \quad . \tag{5.3.13}$$

Since (Appendix A) $\bar{\sigma}[P^{-1}(j\omega)] = 1/\underline{\sigma}[P(j\omega)]$ it follows that requiring small sensitivity (equivalently, requiring large loop gain) at frequencies for which the plant has small gain in some direction may cause unacceptably large plant inputs in that direction. This tradeoff is analogous to that in the scalar case only when the plant has condition number approximately equal to one. Otherwise, the tradeoff is only present in the directions for which plant gain and sensitivity are both small. Hence, if size of the plant input signal is of concern, then the condition $\kappa[P(j\omega)] \approx 1$ must be satisfied in order that the multivariable system behave like its scalar counterpart.

Analytic tradeoffs, between properties of multivariable systems in different frequency ranges, are also present, although not at all well-understood. We shall next discuss some generalizations of the integral relations which quantify these tradeoffs in scalar systems and explain why the generalizations are primarily useful only for the special class of problems discussed in this section.

Our first result is a generalization of the tradeoff imposed by the Bode sensitivity integral discussed in Section 3.4.

Theorem 5.3.1: Assume that the open loop transfer function $L_O(s)$ has entries which are rational functions with at least two more poles than zeros. Denote the poles of $L_O(s)$ in the open right half plane by $\{p_i\ ;\ i = 1, \ldots, N_p\}$, including multiplicities. Then, if the closed loop system is stable, the determinant of the sensitivity function must satisfy

$$\int_0^\infty \log|\det[S_O(j\omega)]|\,d\omega = \pi \sum_{i=1}^{N_p} \mathrm{Re}[p_i] \tag{5.3.14}$$

■

Proof: See Appendix C . ■

Corollary 5.3.2: Assume that the hypotheses of Theorem 5.3.1 are satisfied. Then the singular values of the sensitivity function satisfy

$$\sum_{i=1}^{n}\int_{0}^{\infty}\log\sigma_i[S_O(j\omega)]d\omega = \pi\sum_{i=1}^{N_p}\mathrm{Re}[p_i] \quad . \tag{5.3.15}$$

■

The corollary shows that a tradeoff between sensitivity reduction and sensitivity increase exists for multivariable systems. The key difference is that the tradeoff holds for the *sum* of the log magnitudes of the singular values. This fact suggests that it might be possible to trade off sensitivity properties in different directions as well as in different frequency ranges. Indeed, it follows easily from (5.3.7-8) and (5.3.15) that

$$\int_{0}^{\infty}\log\underline{\sigma}[S(j\omega)] \le \frac{\pi}{n}\sum_{i=1}^{N_p}\mathrm{Re}[p_i] \tag{5.3.16}$$

and[12]

$$\int_{0}^{\infty}\log\bar{\sigma}[S(j\omega)]d\omega \ge \frac{\pi}{n}\sum_{i=1}^{N_p}\mathrm{Re}[p_i] \tag{5.3.17}$$

It does not appear easy to make any stronger general statements about the values of these integrals. Presumably, therefore, one could require that $\underline{\sigma}[S_O(j\omega)] < 1 \;\forall\; \omega$, so that the benefits of feedback in reducing sensitivity could be obtained in the direction of the associated singular subspaces at all frequencies. The integral (5.3.15) shows, however, that a tradeoff between sensitivity increase and sensitivity reduction must still be performed for the other singular values. Indeed, the tradeoff for these singular values will be even more severe, since

---

[12]This is a slight strengthening of a result due to Boyd and Desoer [BoD84]. These authors show that $\int_{0}^{\infty}\log\bar{\sigma}[S(j\omega)]d\omega \ge 0$.

the net area of sensitivity reduction associated with $\underline{\sigma}[S_O(j\omega)]$ must be compensated by an additional area of sensitivity increase for the other singular values.

For the unstructured uncertainty, noise, and disturbances we are considering, design requirements will dictate that S_O and T_O have approximately equal levels of gain at all frequencies. Hence the singular values of S_O should satisfy $\sigma_i[S_O(j\omega)] \approx \sigma_j[S_O(j\omega)]\ \forall\ i,j$ and $\forall\ \omega$. In this case, the tradeoff imposed by the integral (5.3.15) is essentially the same as for scalar systems. Since feedback properties are desired to be the same in all directions, there is no advantage to be gained by attempting to make tradeoffs among the singular values and subspaces.

A tradeoff analogous to that imposed by the Bode gain-phase relation is also present in multivariable system design. This result was first shown by Doyle [DoS81] (see also [HuM82]). The underlying ideas are that the eigenvalues of a rational matrix $M(s)$ must satisfy the bounds $\underline{\sigma}[M(s)] \leq |\lambda_i[M(s)]| \leq \overline{\sigma}[M(s)]$ and are also *algebraic* functions [PoM79] of the complex frequency variable. The former fact implies that

$$\underline{\sigma}[L_O(j\omega)] \leq |\lambda_i[L_O(j\omega)] \leq \overline{\sigma}[L_O(j\omega)] \qquad (5.3.18)$$

and

$$\overline{\sigma}[S_O(j\omega)] \geq 1/|1+\lambda_i[L_O(j\omega)]| \quad . \qquad (5.3.19)$$

The latter fact implies that the eigenvalues must satisfy an integral relation analogous to the scalar gain-phase relation (although one must insure that the effect of right half plane branch points is correctly accounted for [DoS81]). Suppose that the open loop singular values are required to roll off at the same rate and to cross over at the same frequency. Then this integral relation, together with (5.3.18-19), may be used to show that a limitation on rate of gain decrease near crossover exists and is essentially identical to that present in a scalar system.

When design requirements specify that the open loop transfer function have different levels of gain in different directions, then a spread in open loop singular values will exist. This property implies, in turn, that the behavior of the singular values need not even roughly approximate that of the eigenvalue gains. Hence, presumably, some of the open loop singular values could roll off rapidly near crossover, while the magnitudes of the eigenvalues decrease at a much slower rate. In such situations, the gain-phase relations satisfied by the eigenvalues convey less useful insight into the associated design tradeoffs.

The integral constraints imposed by open right half plane poles and zeros may also be generalized to multivariable systems. For example, Boyd and Desoer [BoD84] show that if $L_O(s)$ has an open right half plane zero, $z = x + jy$, then

$$\int_0^\infty \log\overline{\sigma}[S_O(j\omega)]\, W(z,\omega) d\omega \geq 0 \quad . \tag{5.3.20}$$

Note the similarities (and differences) between this result and the integral (3.2.16); in particular, no information about the tightness of the bound is currently available. It is interesting to note that the mathematical tool used in [BoD84] is the fact that $\log\overline{\sigma}[S_O(j\omega)]$ is a *subharmonic* function and must satisfy Poisson's *Inequality*. The reader is referred to [BoD84] for additional results on tradeoffs in multivariable systems; see also Section VII of[FrL85a].

To summarize, we have seen that straightforward generalizations of scalar system concepts are available; however, they are useful mostly in special cases where a multivariable system behaves sufficiently like a scalar system. In particular, we had to assume that the plant transfer function was well-conditioned and that uncertainty, noise, and disturbances are all unstructured. Relaxing the former assumption and studying problems associated with ill-conditioned plants is beyond the scope of this monograph. We shall, instead, devote our attention to the study of system properties in the case that uncertainty, disturbances, and noise are not unstructured and thus are significantly larger in some directions than in others.

5.4. Structure and Multivariable Design Specifications

We saw in the preceding section that if uncertainty, disturbances, and noise are unstructured, then design goals can be quantified by simple closed-loop design specifications such as (5.3.5). In multivariable systems, however, it is reasonable to expect that there will exist frequency ranges over which the levels of uncertainty, disturbances, and noise are significantly different in different directions. The existence of such structure implies that design goals can no longer be adequately quantified by simple bounds. As a consequence, the rules of thumb (5.3.10-11) relating open and closed loop system properties no longer suffice to reveal how a compensator should be chosen. Furthermore, the algebraic and analytic design tradeoffs no longer manifest themselves in a form readily analogous to the scalar case. Our purpose in this section is to discuss some examples for which the existence of structure prevents simple extensions of scalar analysis methods from being applicable.

Consider first the tradeoff between disturbance response and noise response governed by the identity (5.3.6). Whenever disturbances and noise are unstructured, and outputs are equally important, this tradeoff is essentially identical to that for scalar systems. In particular, the sensitivity and complementary sensitivity functions should each have approximately equal gains in all directions; hence the rules of thumb (5.3.10-11) call for open loop gain to have approximately the same level in all directions at each frequency.

Suppose instead that sensors in some loops of the system become noisy at a lower frequency than do sensors in the other loops. This situation can be modelled mathematically by assuming that the sensor noise can be written as the sum of a structured and an unstructured component:

$$n(s) = n_s(s) + n_u(s) \tag{5.4.1}$$

The structured noise n_s is assumed to lie in a k-dimensional subspace $\mathbf{N}_s \subseteq \mathbb{C}^n$, and satisfy the bound

$$||n_s(j\omega)|| \le M_{n_s}(\omega) \quad , \quad \forall\ \omega \ . \tag{5.4.2}$$

The remaining unstructured component of the noise satisfies the bound

$$||n_u(j\omega)|| \le M_{n_u}(\omega) \quad , \quad \forall\ \omega \ . \tag{5.4.3}$$

We shall continue to assume that the disturbance is unstructured and is bounded by $M_d(\omega)$ (5.3.1) and that the response to noise and disturbances must be bounded by (5.3.3) and (5.3.4).

We are assuming that the level of the *structured* component of the noise becomes larger than that of the disturbance at a lower frequency than does the level of the unstructured component. This can be modelled by supposing that the ratios $M_{n_s}(\omega)/M_d(\omega)$ and $M_{n_u}(\omega)/M_d(\omega)$ are as depicted in Figure 5.4.1. Let the orthogonal projections onto $\mathbf{N}_s$, the subspace which contains the structured noise, and onto $\mathbf{N}_s^\perp$, its orthogonal complement, be denoted by P_s and $P_s^\perp$, respectively. We can use these projections, together with the information supplied by Figure 5.4.1, to show that the design specification for this system should satisfy

$$\bar{\sigma}[S_O(j\omega)] \ll 1 \quad \omega \le \omega_1 \tag{5.4.4a}$$

$$\bar{\sigma}[T_O(j\omega)P_s] \ll 1 \quad \omega \ge \omega_2 \tag{5.4.4b}$$

$$\bar{\sigma}[S_O(j\omega)P_s^\perp] \ll 1 \quad \omega \le \omega_3 \tag{5.4.4c}$$

$$\bar{\sigma}[T_O(j\omega)] \ll 1 \quad \omega \ge \omega_4 \tag{5.4.4d}$$

$$\bar{\sigma}[S_O(j\omega)] \le M_S(\omega) \quad \forall\ \omega \tag{5.4.4e}$$

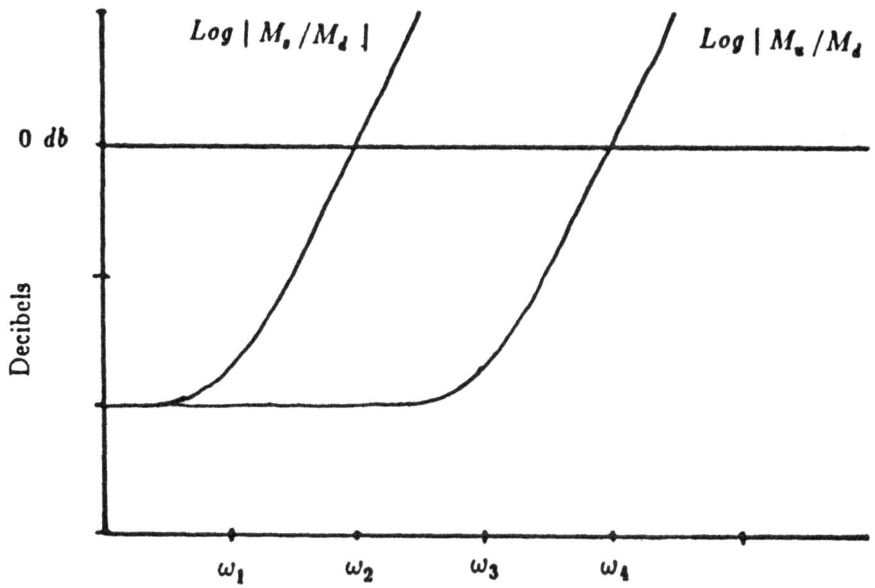

Figure 5.4.1: Relative Levels of Sensor Noise to Disturbance Inputs

$$\bar{\sigma}[T_O(j\omega)] \le M_T(\omega) \quad \forall\ \omega \quad . \tag{5.4.4f}$$

At low frequencies, where disturbances dominate in all directions, (5.4.4a) requires that sensitivity be small in all directions. At frequencies for which the structured noise dominates the disturbance, the complementary sensitivity function should be small in the direction of the structured noise (5.4.4b). Sensitivity, on the other hand, should remain small in the direction of the unstructured noise over the frequency range in which disturbances dominate (5.4.4c).[13] At high frequencies, where sensor noise dominates in all directions, the complementary sensitivity function should be small everywhere (5.4.4d). Finally, the bounds (5.4.4e-f) require that sensitivity and complementary sensitivity not be excessively large in any direction at any frequency.

Notice that this design specification requires the existence of a frequency range (ω_2,ω_3) over which both S_O and T_O are small, but in different directions. This specification is consistent with the design tradeoff imposed by (5.3.6) since this equation only requires that a tradeoff be made between the responses to noise and disturbances lying in the same direction.

Since the rules of thumb (5.3.10-11) only show how to make $\bar{\sigma}[S_O]$ or $\bar{\sigma}[T_O]$ small in *all* directions, they are not useful in shaping the open loop transfer function so that the specification (5.4.4) is satisfied. Intuition suggests that a specification such as (5.4.4) could be satisfied by requiring loop gain to be (a) large in directions for which sensitivity is required to be small and (b) small in directions for which complementary sensitivity is required to be small. In Chapter 7 we shall show that, under certain conditions, this intuition turns out to be correct.

---

[13]As we shall see in Section 7.6, there are some subtleties associated with this specification; it is possible, at least in principle, to reject disturbances in directions other than $\mathbf{N}_r^{\perp}$.

The presence of structured uncertainty can also yield design specifications requiring the closed loop transfer functions to have different levels of gain in different directions. To see why this is so, suppose that multiplicative uncertainty exists at the plant output and has the form

$$\Delta_0(s) = Y\Delta_s(s)X^T + \Delta_u(s) \quad , \tag{5.4.5}$$

where $Y \in IR^{n \times r}$ and $X \in IR^{n \times q}$ are constant matrices with orthonormal columns, $\Delta_s(s) \in \mathbb{C}^{r \times q}$ is an unstructured perturbation satisfying the bound

$$\bar{\sigma}[\Delta_s(j\omega)] \le M_s(\omega) \quad , \quad \forall \; \omega \quad , \tag{5.4.6}$$

and the perturbation $\Delta_u(s) \in \mathbb{C}^{n \times n}$ represents additional unstructured uncertainty bounded by

$$\bar{\sigma}[\Delta_u(j\omega)] \le M_u(\omega) \quad , \quad \forall \; \omega \quad . \tag{5.4.7}$$

The first term in (5.4.5) represents uncertainty which only affects the system in the directions dictated by the matrices X and Y. For example, if this term arises from unmodelled sensor dynamics in the ith loop, then Y and X are column vectors whose entries are all zero except for a "1" in the ith position. Suppose that a frequency range exists in which uncertainty is significantly large in some (fixed) directions, but remains negligible in others. We can model this situation with the decomposition (5.4.5) by using the matrices X and Y to describe the directions of the uncertainty and adjusting the size of the bounds $M_s(\omega)$ and $M_u(\omega)$ according to the size of the two components of uncertainty.

Following the proof of the stability robustness theorem in [DoS81], it is easy to show that the system is robustly stable if it is stable for $\Delta_O = 0$ and if

$$\det[I + T_O(j\omega)\Delta_O(j\omega)] \neq 0 \quad \forall \; \omega \; , \quad \forall \; \Delta_O \quad . \tag{5.4.8}$$

At frequencies for which the unstructured uncertainty is negligible, (5.4.8) reduces to

$$\det[I + T_O(j\omega)Y\Delta_s(j\omega)X^T] = \det[I + X^T T_O(j\omega)Y\Delta_s(j\omega)] \neq 0 \quad . \tag{5.4.9}$$

The latter condition is satisfied if and only if

$$\bar{\sigma}[X^T T_O(j\omega)Y] < 1/M_s(\omega) \quad . \tag{5.4.10}$$

From this analysis, we see that the presence of structured uncertainty leads to the requirement that the complementary sensitivity function be small, but *only* in directions dictated by the uncertainty structure.

Clearly, one way to satisfy (5.4.10) and insure stability robustness is to require that the open loop gain be small in all directions. This strategy, however, would also imply that $S_O \approx I$ and thus (needlessly) prevent us from achieving the benefits of disturbance rejection and sensitivity reduction in directions not affected by the uncertainty. As in the structured sensor noise problem, intuition suggests that we require loop gain to be small only in the direction of the uncertainty and maintain large loop gain in other directions. We shall return to this problem in Chapter 7.

5.5. Summary

We have seen that in certain special cases scalar analysis methods extend readily to multivariable systems. Essentially, the multivariable system must behave sufficiently like a scalar system so that no uniquely multivariable system phenomena can exist. Unfortunately, this rather strong assumption excludes many design problems arising in applications which do exhibit inherently multivariable properties. Hence there is a strong need to develop additional analysis methods to cope with the effects of directionality on multivariable systems.

One uniquely multivariable system phenomenon arises when uncertainty and disturbances are present at different loop locations and the plant transfer function has a high condition number. Discussion of this problem is beyond the scope of this monograph; the reader is referred to [DWS82, Ste85, FrL85d, FrL86a].

The remainder of this monograph will be devoted to analysis of systems with structured uncertainty, noise, and disturbances. We shall focus on two general types of problems. One of these is understanding the relation between open and closed loop properties of a multivariable system when the special cases (5.3.10-11) fail to apply. This task will be undertaken in Chapter 7.

The other problem we shall consider is how to quantify analytic tradeoffs between feedback properties of multivariable systems in different frequency ranges. We have seen that existing multivariable extensions of the integral relations quantifying these tradeoffs in scalar systems are useful only in special cases. In Chapters 8-12 we shall develop a set of mathematical tools with which analytic tradeoffs can be analyzed in an inherently multivariable setting.

Finally, we note that the role of phase in multivariable systems is far from being well-understood. Nevertheless, we shall see that various notions of phase appear to play an essential role in the theory to be developed in Chapters 7-12. Hence, before proceeding to these chapters, we first devote Chapter 6 to studying what role, if any, is played by phase in a multivariable system.

CHAPTER 6

GAIN, PHASE, AND DIRECTIONS IN MULTIVARIABLE SYSTEMS

6.1. Introduction and Motivation

Classical loop-shaping design and analysis methods focused on manipulating the gain and phase of a scalar transfer function. The important role that these coordinates played in classical design has provided considerable motivation for generalizing the concepts of gain and phase to multivariable systems. Such extensions are not completely straightforward, however, due to the directionality properties that multivariable systems generally possess. As we saw in Chapter 5, directionality implies that matrix transfer functions can possess quite different properties, such as the value of gain, in different directions over the same frequency range. This phenomenon must be taken into account if multivariable extensions of gain and phase are to be developed.

It is relatively straightforward to generalize the concept of gain to multivariable systems by using the singular values of a matrix transfer function ([Doy79, MaS79]). As discussed in Section 5.3, the singular values of a matrix may be thought of as a canonical set of gains, including the largest and smallest possible gains at a given frequency.

Associated with the singular values is a set of left and right singular vectors which contain directionality information. Specifically, these vectors determine in what directions the system has the level of gain given by a particular singular value. Although directionality properties are by no means well-understood, the singular vectors have been successfully used in several analyses of feedback system properties (e.g.,[LCL84, FLC82]). At the least, the idea that matrix transfer functions and vector signals exhibit directionality properties is generally accepted.

Extending the concept of scalar phase to multivariable systems has not been at all straightforward. On the one hand, the importance of the very concept of phase has been called into question. In particular, it has been argued (e.g.,[LCL81, AsH84]) that the classical notion of phase margin may not be of primary importance even in the case of scalar systems. On the other hand, several different definitions of multivariable phase have been proposed (see, e.g., the work of MacFarlane and co-workers[PEM81, MaH81, HuM82]). Given these observations, perhaps the most accurate assessment is that a generally accepted concept of phase in multivariable systems is not yet available. Nevertheless, the idea of such a generalization seems worth pursuing. For example, it has been shown [WDH80] that transmission zeros of multivariable systems contribute extra phase lag in some directions, but not in others. Certainly, it would be desirable to have a more precise description of this behavior.

Our purpose in this monograph is not to attempt a definitive generalization of phase to multivariable systems. Rather, we shall explore some physical interpretations of phase in scalar systems and then see if quantities with analogous properties exist in multivariable systems. Toward this end, we note that phase played several different roles in classical feedback theory. First, phase may be viewed merely as one of the two real coordinates needed to describe a complex-valued transfer function. We shall see, in Section 9.5, that phase-like coordinates also arise in parameterizing multivariable transfer functions.

More significantly, the value of phase near gain crossover frequency is critical to maintaining a satisfactory stability margin. In Chapter 7 we shall see that the idea of gain crossover frequency extends, at least in principle, to multivariable systems. Furthermore, in special cases of interest stability margins in these crossover regions depend upon a generalized notion of phase.

As we discussed in Chapter 2, phase is also the property of a stable, minimum-phase, scalar transfer function whose value is completely determined once the gain has been specified. This is an important property since, loosely speaking, it shows that the two real coordinates needed to describe a scalar transfer function yield only one degree of freedom to be manipulated in design. Intuitively, it would seem reasonable that specifying one singular value of a matrix transfer function would determine the value of some other coordinate needed to describe the matrix. In Chapters 8-11, we shall see that this is indeed the case, once appropriate directionality properties are taken into account.

The purpose of this chapter is to show how one additional property of scalar phase can be extended to a multivariable setting. Specifically, we note that the phase *difference* between a pair of scalar signals determines how the signals interfere when added together, and take this role of phase to be of primary physical significance. In Section 6.2 we shall review the role played by phase difference in determining how scalar signals interfere when added. A quantity playing a similar role exists in the multivariable case as well, and will be discussed in Section 6.3. In Section 6.4 this notion of phase difference is related to the numerical range of a matrix transfer function. Section 6.5 discusses a method of assigning a measure of phase difference to each singular value. Section 6.6 contains a summary and conclusions.

6.2. Phase in Scalar Systems

Consider two nonzero scalar exponential signals with the same value of complex frequency, $s_0 = x_0 + jy_0$:

$$w_i(t) = c_i e^{s_0 t} \qquad i = 1, 2 \ . \tag{6.2.1}$$

The phase difference between these signals may be uniquely defined (modulo 2π) as

$$\begin{aligned}\Delta\theta(w_1,w_2) &= \measuredangle\, \overline{w}_1(t)w_2(t) \\ &= \measuredangle\, \overline{c}_1\, c_2\end{aligned} \qquad (6.2.2)$$

The physical significance of the phase difference between these signals is that it governs how the signals interfere when added together:[14]

$$|w_1(t) + w_2(t)| = e^{s_0 t}\sqrt{|c_1|^2+|c_2|^2+2\mathrm{Re}[\overline{c}_2 c_1]} \qquad (6.2.3a)$$

$$= e^{s_0 t}\sqrt{|c_1|^2+|c_2|^2+2|c_1|\cdot|c_2|\cos\Delta\theta}\ . \qquad (6.2.3b)$$

For example, if $\Delta\theta = (2k+1)\pi$, k an integer, then

$$|w_1(t) + w_2(t)| = e^{s_0 t}\Big|\,|c_1|-|c_2|\,\Big| \qquad (6.2.4)$$

and the two signals interfere purely destructively. If $\Delta\theta = 2k\pi$, then

$$|w_1(t) + w_2(t)| = e^{s_0 t}[\,|c_1| + |c_2|\,] \qquad (6.2.5)$$

and the two signals interfere purely constructively. If the phase difference is not an integer multiple of π, then the magnitude of the sum of the two signals is bounded below and above by (6.2.4) and (6.2.5).

The preceding remarks may be used to give a useful interpretation to the phase of a scalar transfer function. Consider a linear time-invariant system with transfer function $L(s)$ and input signal $u(t) = ce^{s_0 t}$ with s_0 not a pole of $L(s)$. It can be shown [DeS74] that the input may be augmented with a set of delta functions and derivatives of delta functions so that the steady state output is reached at $t = 0^+$. The time response of the output is then related to

---

[14] Throughout this section, arguments of $\Delta\theta(\cdot,\cdot)$ will be suppressed when notationally convenient.

that of the input via the transfer function:

$$y(t) = L(s_0)u(t) \quad , \quad t > 0 \ . \tag{6.2.6}$$

Hence, for scalar systems, the phase difference between input and output is determined by the phase of the transfer function:

$$\begin{aligned} \Delta\theta(u,y) &= \sphericalangle\, \overline{u}(t)y(t) \\ &= \sphericalangle\, \overline{u}(t)L(s_0)u(t) \\ &= \sphericalangle\, L(s_0) \end{aligned} \tag{6.2.7}$$

From (6.2.3), it follows that the phase of a transfer function determines how the input and output will interfere when added together. The magnitude of the sum of these two signals is

$$|u(t) + y(t)| = e^{x_0 t}\, |c| \cdot \left|1+L(s_0)\right| \tag{6.2.8a}$$

$$= e^{x_0 t}\, |c| \cdot \sqrt{1+|L(s_0)|^2+2|L(s_0)|\cos \sphericalangle\, L(s_0)} \tag{6.2.8b}$$

Phase difference has an obvious relation to the feedback properties of a system. Suppose a (negative) feedback loop is closed around a system with transfer function $L(s)$. It is well known that the return difference $1+L(s)$ and its reciprocal, the system sensitivity function, express important properties of a feedback system. Equation (6.2.8a) explicitly relates these feedback properties to the way in which inputs $u(t) = ce^{s_0 t}$ and the resulting outputs interfere when added together. Equation (6.2.8b) shows that $\sphericalangle\, L(s_0)$ determines whether this interference is more nearly constructive or destructive.

In the next two sections the preceding well-known observations will be used to investigate phase and phase difference in multivariable systems.

6.3. Phase Difference between Vector Signals

The interpretation of phase difference in scalar systems reviewed in Section 6.2 suggests that a reasonable approach to phase in multivariable systems is through study of how two vector signals interfere when added together. Consider two complex exponential signals taking values in $\mathbb{C}^n$:

$$w_i(t) = c_i e^{s_o t} \quad , \quad c_i \in \mathbb{C}^n \tag{6.3.1}$$

The analogue to (6.2.3a) is[15]

$$\|w_1(t) + w_2(t)\| = e^{s_o t}\sqrt{\|c_1\|^2 + \|c_2\|^2 + 2\mathrm{Re}[c_1^H c_2]} \ , \tag{6.3.2}$$

where c_1^H is the complex conjugate of the transpose of c_1. Equation (6.3.2) shows that the interference of the two signals is governed by the term $\mathrm{Re}[c_1^H c_2]$. This quantity will now be investigated in detail.

An important difference between scalar and multivariable systems is that signals in the latter are vector-valued. Hence each signal lies in a particular *direction*, or one-dimensional subspace of $\mathbb{C}^n$. As we shall see, directionality plays an important role in the generalization of phase difference to multivariable signals and systems. Before proceeding, we need to adopt some notational conventions for describing directions in $\mathbb{C}^n$. First, the phrase "the direction $\mathbf{C}$" means "the one-dimensional subspace $\mathbf{C} \subseteq \mathbb{C}^n$." Second, when we are given a vector c, the direction in which it lies will be denoted $[c]$.

Given two vector-valued signals, it is natural to ask how close together they lie. To study this, a measure of the angular distance between two directions will be introduced. Consider two one-dimensional subspaces $\mathbf{C}_1, \mathbf{C}_2 \subseteq \mathbb{C}^n$. It is standard to define the angle between

---

[15] As elsewhere in this monograph, the standard Euclidean vector norm and induced matrix two-norm will be used exclusively.

these subspaces as

$$\phi(C_1 , C_2) \triangleq arccos\left\{ \frac{|c_1^H c_2|}{||c_1|| \cdot ||c_2||} \right\} \tag{6.3.3}$$

where $c_i \in C_i$ are *arbitrarily* chosen nonzero vectors. Note that $0 \le \phi \le \pi/2$, and that ϕ is an angular measure of distance between the subspaces C_1 and C_2. If $\phi = 0$, the subspaces are identical; if $\phi = \pi/2$, the subspaces are orthogonal. Given two signals $w_i = c_i e^{s_0 t}$, the angular distance between them can be defined from (6.3.3) via the formula $\phi(w_1,w_2) = \phi([c_1],[c_2])$.

Suppose we are given two subspaces C_1 and C_2 which are not orthogonal. Then the phase difference between two signals $w_i(t) = c_i e^{s_0 t}$, with $c_i \in C_i$, may be defined as

$$\begin{aligned} \Delta\theta(w_1,w_2) &\triangleq \measuredangle\, w_1^H(t) w_2(t) \\ &= \measuredangle\, c_1^H c_2 \end{aligned} \tag{6.3.4}$$

Since $\Delta\theta(w_1,w_2)$ depends only upon the vectors c_1 and c_2, we may write $\Delta\theta(w_1,w_2) = \Delta\theta(c_1,c_2)$. Using (6.3.3) and (6.3.4) in (6.3.2) yields[16]

$$||w_1(t) + w_2(t)|| = e^{x_0 t}\sqrt{||c_1||^2+||c_2||^2+2||c_1||\cdot||c_2||\cos\phi\cos\Delta\theta}\ . \tag{6.3.5}$$

Comparing (6.3.5) with (6.2.3b) reveals the role played by directionality in determining how the two signals interfere. Indeed, this role is quantified by the angle ϕ, which determines the extent to which the signals *can* interfere when added together. If $\phi = 0$, so that the signals lie in the same direction, the phase difference plays precisely the same role as it does in

[16]In this section, the arguments of $\phi(\cdot,\cdot)$ and $\Delta\theta(\cdot,\cdot)$ will be suppressed when notationally convenient.

scalar systems. That is, $\Delta\theta = (2k+1)\pi$, k an integer, implies destructive interference:

$$\| w_1(t) + w_2(t) \| = e^{x_0 t} \big| \| c_1 \| - \| c_2 \| \big| \ , \tag{6.3.6}$$

while $\Delta\theta = 2k\pi$ implies constructive interference:

$$\| w_1(t) + w_2(t) \| = e^{x_0 t} [\| c_1 \| + \| c_2 \|] \ . \tag{6.3.7}$$

If $0 < \phi < \pi/2$, then only that component of $w_2(t)$ in the direction of $w_1(t)$ can cause interference when the two signals are added together. Again, the phase difference $\Delta\theta$ determines the type of interference which does take place. Finally, if $\phi = \pi/2$, the phase difference (6.3.4) is undefined. Physically, this is quite reasonable, since $\phi = \pi/2$ implies the two signals lie in orthogonal subspaces and thus *cannot* interfere when added together. In this case the magnitude of the sum of the two signals is given by the Pythagorean Theorem:

$$\begin{aligned} \| w_1(t) + w_2(t) \| &= \sqrt{\| w_1(t) \|^2 + \| w_2(t) \|^2} \\ &= e^{x_0 t} \sqrt{\| c_1 \|^2 + \| c_2 \|^2} \ . \end{aligned} \tag{6.3.8}$$

It is interesting to note that (6.3.8) also holds for signals lying in the same direction (i.e., $\phi(w_1, w_2) = 0$) whose phase difference is $\Delta\theta(w_1, w_2) = (2k+1)\pi/2$.

6.4. Phase Difference and Matrix Transfer Functions

Consider a linear time invariant system with transfer function matrix $L(s)$ taking values in $\mathcal{C}^{n \times n}$. As in the scalar case, the output of this system due to a (vector) input augmented with appropriate delta functions and their derivatives is given by (6.2.6), where $u(t)$ and $y(t)$ take values in $\mathcal{C}^n$. Clearly, the action of the system upon an input signal is a function of the direction of the input. Consider an input signal in the direction C: $u(t) = ce^{s_0 t}$, $s_0 = x_0 + jy_0$, where $c \in \mathrm{C}$ is an arbitrarily chosen *nonzero* vector. To simplify the notation, we shall also assume that c has *unit* magnitude.

When referring to the gain of a multivariable system, it is necessary to specify both the direction as well as the frequency of interest. Hence we define the gain of $L(s)$ at frequency s_0 and in the direction C by

$$\rho(s_0,\mathrm{C}) \triangleq \frac{\| L(s_0)u(t) \|}{\| u(t) \|} = \| L(s_0)c \| \tag{6.4.1}$$

Note that the assumption $\| c \| = 1$ has been used in definition (6.4.1).

The angle between an input $u(t)$ lying in the direction C and the resulting output $y(t)$ can be computed from (6.3.3) and will be denoted

$$\begin{aligned} \phi(s_0,\mathrm{C}) &\triangleq \phi(u(t),y(t)) \\ &= \arccos \left[\frac{| u^H(t)y(t) |}{\| u(t) \| \cdot \| y(t) \|} \right] \\ &= \arccos \left[\frac{| c^H L(s_0)c |}{\rho(s_0,\mathrm{C})} \right] . \end{aligned} \tag{6.4.2}$$

The phase difference between an input in the direction C and the resulting output can be computed from (6.3.4) whenever $\rho \neq 0$ and $\phi \neq \pi/2$. We shall denote this parameter by

$$\begin{aligned} \Delta\theta(s_0,\mathrm{C}) &\triangleq \Delta\theta(u(t),y(t)) \\ &= \measuredangle\, u^H(t)y(t) \\ &= \measuredangle\, e^{\bar{s}_0 t} c^H L(s_0)c e^{s_0 t} \\ &= \measuredangle\, c^H L(s_0)c \end{aligned} \tag{6.4.3}$$

Consider an input of the form $ce^{s_0 t}$, where c is a unit vector. The gain, the angle between the subspaces spanned by the input and output, and the phase difference between the input and output are all related. This is readily seen by factoring the term

$$c^H L(s_0)c \equiv \rho(s_0,\mathrm{C})\cos\phi(s_0,\mathrm{C})\exp[j\Delta\theta(s_0,[c])] \quad . \tag{6.4.4}$$

Denote the set of all such terms, obtained by letting c range over the unit sphere in $\mathbb{C}^n$, by

$$\mathrm{L}(s_0) \triangleq \{c^H L(s_0)c \ : \ \|c\|=1\} \tag{6.4.5}$$

The set $\mathrm{L}(s_0)$ is the *numerical range* [Hal58] of $L(s_0)$, and has previously been identified as having importance in system theory. For example, a necessary condition for strict positive realness [And68] of the transfer function $L(s)$ is that the matrix $L(s) + [L(s)]^H$ be positive definite $\forall$ s with $\mathrm{Re}[s] \geq 0$. From $c^H[L(s)+[L(s)]^H]c = 2\mathrm{Re}[c^H L(s)c]$ it follows that this condition is equivalent to requiring that the set $\{\mathrm{L}(s_0) | \mathrm{Re}[s_0] \geq 0\}$ be confined to the open right half plane. From (6.4.4) the latter condition is satisfied if and only if $\rho(s_0,\mathrm{C}) > 0$, $\phi(s_0,\mathrm{C}) < \frac{\pi}{2}$ and $|\Delta\theta(s_0,\mathrm{C})| < \frac{\pi}{2}$ $\forall$ s with $\mathrm{Re}[s] \geq 0$ and $\forall$ $c \neq 0$.

From (6.3.5), it follows that the magnitude of the sum of an input signal $u(t) = ce^{s_0 t}$ and the resulting output is given by

$$\begin{aligned} \|u(t)+y(t)\| &= \|[I+L(s_0)]c\| \\ &= e^{x_0 t}\sqrt{1+\rho^2+2\rho\cos\phi\cos\Delta\theta} \quad . \end{aligned} \tag{6.4.6}$$

(In (6.4.6) we have suppressed the dependence of ρ, ϕ, and $\Delta\theta$ upon s_0 and C for sake of notational simplicity). Just as in the scalar case, the appearance of the return difference in (6.4.6) implies that the parameters ρ, ϕ, and $\Delta\theta$ contain information about properties of a feedback system whose open loop transfer function is $L(s)$. The feedback properties are clearly related to the way in which an input to the system described by $L(s)$ and the resulting output interfere when added together. Unlike the scalar case, the parameters ρ and $\Delta\theta$, which generalize the concepts of gain and phase, do not suffice to describe the interference between the two signals. It is also necessary to consider the additional parameter, ϕ, quantifying the misalignment of the input and output directions due to a spatial rotation of the signal by the system.

Equation (6.4.6) only yields information about the size of the return difference matrix in one direction. The smallest possible size of this matrix may be obtained from the definition of the smallest singular value:

$$\underline{\sigma}[I+L(s_0)] \triangleq \inf_{\|c\|=1} \| [I+L(s_0)]c \| \quad . \tag{6.4.7}$$

In the present context, taking the infimum in (6.4.7) may be interpreted as finding the input direction (or directions) at frequency s_0 for which the resulting output most nearly cancels the input when the two signals are added together.

The ideas discussed above can be useful in analyzing the relation between open and closed loop system properties. For example, suppose that the open loop gain in some direction is equal to unity, $\rho(s_0,C) = 1$, and further suppose that the phase difference in this direction is $\Delta\theta(s_0,C) = \pm 180°$. If the input and output signals span the same subspace ($\phi(s_0,C) = 0$), so that c is an eigenvector of $L(s_0)$, then the return difference matrix is singular at s_0. Otherwise, $\| [I+L(s_0)]c \| > 0$ and (6.4.6-7) may be used to derive conditions on ϕ necessary to insure that $\underline{\sigma}[I+L(s_0)]$ satisfies a given lower bound. In the present example, with $\rho(s_0,C)\exp[j\Delta\theta(s_0,C)] = -1$, it is necessary that $\phi(s_0,C) \geq 60°$ to maintain $\| [I+L(s_0)]c \| \geq 1$. With this condition satisfied, one would have to apply a spatial rotation of at least 60° to the output in order that c be an eigenvector of $L(s_0)$ with eigenvalue -1. This observation may prove useful in studying the structure of feedback designs whose sensitivity functions satisfy some bound; e.g., linear quadratic state feedback designs performed under certain restrictions on the weighting matrices[DoS81].

6.5. The Angle between Singular Subspaces and Phase Difference Between Singular Vectors

The construction of Section 6.4 may be used to associate a measure of phase difference to each singular value of a square matrix transfer function. Recall that to each singular value $\sigma_i \equiv \sigma_i[L(s_0)]$ with multiplicity one there corresponds a pair of uniquely defined one-dimensional left and right singular subspaces, denoted $\mathbf{V}_i$ and $\mathbf{U}_i$, respectively. The right singular vector may be chosen as an arbitrary unit vector $u_i \in \mathbf{U}_i$; if $\sigma_i \neq 0$, the left singular vector is then uniquely determined from the relation $v_i = L(s_0)u_i/\sigma_i$. Provided that $\mathbf{V}_i$ and $\mathbf{U}_i$ are not orthogonal subspaces, (6.4.3) may be used to define the phase difference between the ith pair of singular vectors.

$$\begin{aligned}\Delta\theta(i) &\triangleq \Delta\theta(s_0,\mathbf{U}_i)\\ &= \Delta\theta(u_i,v_i)\\ &= \sphericalangle\, u_i^H L(s_0)u_i\\ &= \sphericalangle\, u_i^H v_i\end{aligned} \tag{6.5.1}$$

Similarly (6.4.2) yields the angle between the ith pair of left and right singular subspaces:

$$\begin{aligned}\phi(i) &\triangleq \phi(s_0,\mathbf{U}_i)\\ &= \phi(\mathbf{U}_i,\mathbf{V}_i)\\ &= arccos\left[\frac{|u_i^H L(s_0)u_i|}{\sigma_i}\right]\\ &= arccos\, |u_i^H v_i| \quad ,\end{aligned} \tag{6.5.2}$$

From (6.4.6-7), the angle between the singular subspaces and the phase difference between the singular vectors can be related to sensitivity properties via the bound

$$\bar{\sigma}[S] \geq 1/\sqrt{1+\sigma_i^2+2\sigma_i\cos\phi(i)\cos\Delta\theta(i)} \quad . \tag{6.5.3}$$

Since (6.5.3) is only a *lower* bound on sensitivity, its usefulness is limited to establishing conditions on the singular vectors which must *necessarily* be satisfied if sensitivity is to be small. In Chapter 7 we shall see that, with relevant additional assumptions, the size of the sensitivity function can actually be approximated using the parameters $\phi(i)$ and $\Delta\theta(i)$.

6.6. Summary

In this chapter we have taken steps toward generalizing one of the many properties of scalar phase to multivariable systems. It is interesting to note that the property which we take to be of primary importance arises from consideration of *signals*, rather than transfer functions of systems. Indeed, we are concerned with the phase *difference* between two signals and how that difference determines the way in which they will interfere when added together. The phase of a scalar transfer function is thus important because it determines the phase difference between the input to and output from a scalar linear system.

We showed in this chapter that there exists a natural definition of phase difference between vector-valued signals. Furthermore, this measure of phase difference determines how the signals will interfere when added together. The new phenomenon arising in the vector case is that the signals will not generally lie in the same direction. Hence the phase difference only determines how the components of the two signals which do lie in the same direction interfere. It is therefore intuitively reasonable that our definition of phase difference is not applicable when the signals lie in orthogonal directions.

It is reasonable to ask whether some notion of the "phase" of a matrix transfer function exists. To answer this question, we showed that the phase of each element of the numerical range of a matrix determines how one pair of input and output signals interfere when added together. That the numerical range arises in our study is perhaps not surprising, since this set has previously arisen in system theory studies. Finally, we closed this chapter by noting that we can use our ideas to measure the phase difference between each pair of singular vectors. As noted in the introduction to this chapter, we shall see multivariable phase arising in several contexts throughout the remainder of this monograph. Of particular relevance to the preceding discussion is a mathematical framework we shall develop in Chapter 9. This framework will allow us to describe, in some detail, both the phase difference between a pair of vector signals as well as the directions in which they lie. We shall show at the close of Chapter 9 that the action of a linear system upon a vector signal can be decomposed into three terms: a change in magnitude, a change in direction, and a change in phase.

CHAPTER 7

THE RELATION BETWEEN OPEN LOOP AND CLOSED LOOP PROPERTIES OF MULTIVARIABLE FEEDBACK SYSTEMS

7.1. Introduction

As we saw in Chapter 2, the relations between open-loop and closed-loop properties of scalar feedback systems are fairly well understood. Furthermore, understanding these relations is important in practice since they immediately yield rules of thumb useful in compensator design. To briefly summarize, at frequencies for which loop gain is sufficiently large, a feedback system has good performance and disturbance rejection properties. Similarly, at frequencies for which loop gain is sufficiently small, a feedback system is robustly stable and rejects sensor noise. If neither of these conditions hold, e.g. at gain crossover, then knowledge of open loop gain is insufficient to allow approximation of feedback properties and phase information is therefore necessary and useful. Note that these relations are useful *despite* the fact that they are only stated as approximations. Indeed, if their only use was to determine feedback properties of a given design, they would be of little interest, since these properties can be *directly* evaluated from certain closed loop transfer functions. The usefulness of these relations stems from several other reasons, however. For example, they may be used to translate closed loop design specifications into specifications on open loop gain and phase. These latter specifications, in turn, are useful in compensator design. The relations are also useful in studying how properties of the open loop system, such as extra phase lag contributed by right half plane zeros, affect feedback properties.

In Chapter 5, we saw that similar relations between open and closed loop properties of multivariable systems hold only in special cases for which gain is either large in all loops or small in all loops. When neither of these assumptions holds, generally useful methods for relating open and closed loop system properties have yet to be developed. Our goal in this chapter is to develop methods applicable to more general cases. We note that our primary motivation is *not* determination of feedback properties for a *given* system, since again this can be done directly from closed loop transfer functions. Rather, we are interested in obtaining insights useful in designing compensators and in studying limitations inherent in any design due to properties of a given plant. Finally, motivated by classical design, we do not demand mathematical precision; we will be satisfied by rules of thumb, approximations of quantitative system properties, and insights into qualitative properties.

The purpose of this chapter is to study the relation between open and closed loop properties of systems which, over some frequency range, have high gains in only some directions,[17] low gains in only some directions, or both high and low gains in different directions. Motivations for studying systems with significantly different levels of gain in different directions are several. First, the physical characteristics of a given plant may dictate that it has different bandwidths in different loops. Second, suppose that the plant has more outputs than inputs. Then, *necessarily*, the open loop transfer function at the plant output will have less than full rank. Similarly, when the plant has more inputs than outputs, the open loop transfer function at the plant input will have less than full rank. In addition, the open loop transfer function will have less than full rank if the gain in some loops is reduced to zero through sensor or actuator failures. Third, if the plant has a right half plane zero, it may be necessary [WDH80] to roll off loop gain in the direction of the zero although large gain can still be maintained in other directions. Finally, as we saw in Section 5.4, the levels of disturbance inputs, sensor noise,

---

[17]By direction we mean a one-dimensional subspace of $\mathbb{C}^n$. The physical loops of the system thus correspond to the directions of the n standard basis vectors.

and system uncertainty in a multivariable system will generally vary with direction as well as with frequency. For example, sensors in some loops may become noisy at a much lower frequency than those in other loops. Hence gain in the loops with noisy sensors must be rolled off before the gain in the other loops, leading naturally to a frequency range over which the open loop transfer function exhibits different levels of gain.

The remainder of this chapter is organized as follows. As we saw in Chapter 5, two sets of transfer functions, equations (5.2.1-3) and (5.2.4-6), are needed to describe the properties of a multivariable feedback system. The reason both sets are needed is that feedback properties are generally different at each loop-breaking point. In this chapter we shall restrict our attention to studying how the open loop transfer function at *one* loop-breaking point is related to feedback properties *at that point*. Hence we drop the subscript "I" or "O" used to distinguish between loop-breaking points, with the understanding that our results are applicable only to transfer functions which are all defined at the same location in the loop. In Section 7.2, the assumption that the open loop transfer function has higher levels of gain in some directions than in others is used to partition its singular value decomposition into interconnected high and low gain "subsystems."[18] The effect of this interconnection upon feedback properties is discussed in Section 7.3, under the assumption that gain in one subsystem is "sufficiently large" while that in the other is "sufficiently small." When one of these assumptions is relaxed, then a situation analogous to gain crossover frequency exists, in that knowledge of open loop gain is insufficient to approximate feedback properties. Hence, in Section 7.4, generalizations of gain crossover for each of the high and low gain subsystems are presented. To illustrate the results an example of an actual design is analyzed in Section 7.5. This example shows how coupling between loops with high and low gains can adversely affect feedback properties.

[18]For a caveat associated with this term, see Section 7.2.

Conclusions and directions for further research are given in Section 7.6.

7.2. Higher and Lower Gain "Subsystems"

In two special cases, classical rules of thumb relating the level of open loop gain to feedback properties generalize readily to multivariable systems (Section 5.3). From the definitions of S and T it follows that if loop gain is large in all directions ($\underline{\sigma}[L(j\omega)] >> 1$) then disturbance response is small in all directions. On the other hand, if loop gain is small in all directions ($\overline{\sigma}[L(j\omega)] << 1$) then response to sensor noise is also small in all directions. These observations are useful in practice since they provide insight into the level of compensator gain required to achieve disturbance or noise rejection.

Suppose, however, that loop gains are neither everywhere large nor everywhere small, perhaps due to one of the reasons cited in Section 7.1. It is then difficult to make any general statement about feedback properties. As we mentioned in Section 5.4, intuition suggests that if gains are sufficiently small in some directions, then response to sensor noise is small in those directions. Similarly, one would expect that if loop gain is large in some directions, then disturbance response is small in those directions. It turns out that this intuition is correct; however, one must distinguish between input directions and output directions as well as consider the effects of coupling between these directions. These notions shall now be made precise.

Motivated by the preceding discussion, we wish to study systems which, over some frequency range, possess much higher levels of gain in some directions than in others. To describe this behavior, it is convenient to use the singular value decomposition of the open loop transfer function matrix L.[19] Recall (Appendix A) that the singular value decomposition

---

[19]Since we are concerned with properties of the system at a single frequency, the complex frequency variable "s" will be suppressed.

of a matrix $M \in \mathbb{C}^{n \times n}$ is given by $M = V \Sigma U^H$, where $\Sigma = diag[\sigma_1 \cdots \sigma_n]$ and $\bar{\sigma} \triangleq \sigma_1 \geq \sigma_2 \geq \cdots \geq \underline{\sigma} \triangleq \sigma_n \geq 0$. The columns of the unitary matrices $V = [v_1 \cdots v_n]$ and $U = [u_1 \cdots u_n]$ are termed the left and right singular vectors, respectively, and satisfy the equation $Mu_i = \sigma_i v_i$. For our purposes, the singular value decomposition of L will be partitioned as

$$L = V_1 \Sigma_1 U_1^H + V_2 \Sigma_2 U_2^H \tag{7.2.1}$$

where $V_1, U_1 \in \mathbb{C}^{n \times k}$, $V_2, U_2 \in \mathbb{C}^{n \times (n-k)}$, $\Sigma_1 = diag[\sigma_1 \cdots \sigma_k]$, and $\Sigma_2 = diag[\sigma_{k+1} \cdots \sigma_n]$. Let the column spaces of the matrices U_i and V_i be denoted $\mathbf{U}_i$ and $\mathbf{V}_i$, respectively. The decomposition (7.2.1) states that inputs to the open loop system lying in the subspace $\mathbf{U}_i$ are amplified by the level of gain represented by the singular values Σ_i and appear as outputs in the subspace $\mathbf{V}_i$. When $\underline{\sigma}[\Sigma_1] >> \bar{\sigma}[\Sigma_2]$, this partition suggests that L be referred to as consisting of a higher gain subsystem ($V_1 \Sigma_1 U_1^H$) and a lower gain subsystem ($V_2 \Sigma_2 U_2^H$) (Caveat: This partition need hold only at the frequency in question and need not correspond to physical subsystems. Strictly speaking, we should say "$\cdots$ subsystem at frequency s".)

The matrices $U_i^H V_j$ give a measure of the alignment of the subspaces $\mathbf{U}_i$ and $\mathbf{V}_j$ and may be interpreted as a measure of interaction between the high and low gain subsystems. To see this, we need the concept of principal angles between a pair of subspaces ([BjG73, GoV83]).[20] Assume that $2k \leq n$. Then the k singular values of the matrix $U_1^H V_1$ are the cosines of the principal angles between the subspaces $\mathbf{U}_1$ and $\mathbf{V}_1$. These angles, which generalize the usual notion of angle when the subspaces are one-dimensional, may be defined

---

[20]These angles have been previously used, not always explicitly, in the control literature. In[BoD84], the largest principle angle is used to study how right half plane poles and zeros interact in multivariable systems. Implicit in the design technique proposed in [HuM82] is that the principal angles should be made small.

recursively from

$$\cos \alpha_j = \max_{u \in \mathbf{U}_1} \max_{v \in \mathbf{V}_1} |u^H v| = \bar{u}_j^H \bar{v}_j \quad ; \; j = 1, \ldots, k \tag{7.2.2}$$

subject to

$$\|u\|_2 = \|v\|_2 = 1$$

$$u^H \bar{u}_i = 0 \qquad i = 1, \ldots, j-1$$

$$v^H \bar{v}_i = 0 \qquad i = 1, \ldots, j-1 \; .$$

The matrix $U_2^H V_2$ has k singular values which are identical to those of $U_1^H V_1$ plus an additional $n - 2k$ singular values which are equal to one.[21] The matrices $U_1^H V_2$ and $U_2^H V_1$ each have k singular values equal to the sines of the principal angles between the subspaces $\mathbf{U}_1$ and $\mathbf{V}_1$. Note that $\frac{\pi}{2} \geq \overline{\alpha} \triangleq \alpha_k \geq \alpha_{k-1} \geq \cdots \geq \underline{\alpha} \triangleq \alpha_1 \geq 0$. If $\alpha_i = 0$, $i = 1, \ldots, k$, then the subspaces $\mathbf{U}_1$ and $\mathbf{V}_1$ are equal; if $\alpha_i = \frac{\pi}{2}$, $i = 1, \ldots, k$, then they are orthogonal and $\mathbf{U}_1 \cap \mathbf{V}_1 = 0$. In terms of systems, the principal angles indicate to what extent the high and low gain subsystems interact when a feedback loop is closed around L. If the principal angles are not all zero, then a component of the output of each subsystem is fed back to the input of the other subsystem. The effect this has on system feedback properties will now be investigated.

7.3. Systems with High and Low Gains at the Same Frequency

In this section we give approximations to the sensitivity and complementary sensitivity functions of a system with both large and small gains in different directions at the same

[21] If $2k > n$, analogous results hold with the roles of $U_1^H V_1$ and $U_2^H V_2$ interchanged.

frequency. Motivated by the discussion in Section 7.1, consider the limiting case in which $\underline{\sigma}[\Sigma_1]$ and $\overline{\sigma}[\Sigma_2]$ in the subsystem decomposition (7.2.1) are allowed to become "sufficiently large" and "sufficiently small", respectively. Intuition suggests that this strategy would achieve sensor noise rejection in an $(n-k)$-dimensional subspace and disturbance rejection in a k-dimensional subspace corresponding to the directions of small and large loop gain, respectively. The following theorem shows that this intuition is correct. However, one must take care to correctly identify the input and output directions of the high and low gain subsystems with the subspaces of $\mathbb{C}^n$ in which the signals are to be rejected.

Theorem 7.3.1: Assume[22] $\det[U_1^H V_1] \neq 0$. Then, if

$$\underline{\sigma}[\Sigma_1] >> \overline{\sigma}[(U_1^H V_1)^{-1}] \tag{7.3.1}$$

and

$$\overline{\sigma}[\Sigma_2] << \underline{\sigma}[U_2^H V_2] \tag{7.3.2}$$

the sensitivity and complementary sensitivity functions satisfy $\overline{\sigma}[S-S_{app}] << 1$ and $\overline{\sigma}[T-T_{app}] << 1$, where

$$S_{app} = U_2(V_2^H U_2)^{-1} V_2^H \tag{7.3.3a}$$

$$= U_2(V_2^H U_2)^{-1}(V_2^H U_1)U_1^H + U_2 U_2^H \tag{7.3.3b}$$

$$= V_1(V_1^H U_2)(V_2^H U_2)^{-1} V_2^H + V_2 V_2^H \tag{7.3.3c}$$

and

$$T_{app} = V_1(U_1^H V_1)^{-1} U_1^H \tag{7.3.4a}$$

---

[22] An analysis of the special case $\det[U_i^H V_i] = 0$ may be performed using the pseudoinverse of $U_i^H V_i$; however, the resulting expressions are complicated.

$$= U_2(U_2^H V_1)(U_1^H V_1)^{-1} U_1^H + U_1 U_1^H \tag{7.3.4b}$$

$$= V_1(U_1^H V_1)^{-1}(U_1^H V_2) V_2^H + V_1 V_1^H \tag{7.3.4c}$$

Proof: See [FrL86b]. ∎

Recall that the subspaces U_1 and V_1 contain, respectively, the inputs to and outputs from the higher gain subsystem, while the subspaces U_2 and V_2 contain the inputs to and outputs from the lower gain subsystem. Equations (7.3.3a) and (7.3.4a) show that the response of the system output to disturbances lying in the subspace V_1 *and* to sensor noise lying in the subspace U_2 can be made arbitrarily small by appropriate choice of open loop gain. With regard to directionality properties, note that the *right* singular subspaces of $L(s)$ determine the direction of the rejected noise, while the *left* singular subspaces determine the direction of the rejected disturbances. Furthermore, note that the *right* singular subspaces determine the subspace of the system output to which the disturbance response is confined, while the *left* singular subspaces determine the subspace of the output to which the noise response is confined.

Observe that $\bar{\sigma}[S_{app}] = \bar{\sigma}[T_{app}] = 1/\cos\bar{\alpha}$, and recall that $\bar{\alpha}$ may be viewed as a measure of coupling between the two subsystems. Hence, to insure that the response of the system output to disturbances lying in the subspace V_2 and to noise lying in the subspace U_1 not be excessively large, it is necessary that the high and low gain subsystems be sufficiently well decoupled. Note that the nonlinear relationship between $\bar{\alpha}$ and $1/\cos\bar{\alpha}$ implies that the effects of coupling do not become pronounced until $\bar{\alpha}$ becomes fairly large (e.g. $\bar{\alpha} = 60° \Rightarrow \bar{\sigma}[S_{app}] \approx 6db$). Hence, although it is necessary to limit the amount of coupling present, insisting on nearly exact alignment of the subspaces U_i and V_i may not be necessary, especially as this goal may conflict with other design objectives.

Further insight may be gained from (7.3.3b-c) and (7.3.4b-c). For example, (7.3.3b) and (7.3.4b) show that the system disturbance and noise responses may each be decomposed into two components. First, disturbances in the subspace U_2 are fed directly to the output of the system (i.e., U_2 is a right invariant subspace of S_{app} with associated eigenvalues equal to one) and noise in the subspace U_2 is rejected. This of course is reasonable since the loop gain is assumed small on this subspace. Hence, as far as both noise and disturbances in U_2 are concerned, the situation is analogous to the single-loop case, for which loop gain being small implies that sensor noise is rejected at the expense of disturbances being fed directly to the system output. Second, both disturbances *and* noise in the subspace U_1 are amplified by the feedback system and produce an output which is confined to the subspace U_2. The amount by which these signals are amplified is determined by the singular values of the matrix $(V_2^H U_2)^{-1}$ $(V_2^H U_1)$; these, in turn, are equal to the *tangents* of the principal angles between the subspaces U_2 and V_2. This second phenomenon, which has no analogue in scalar systems, is due to those components of the output from each subsystem which are fed back to the input of the other subsystem due to coupling. A little thought experiment suggests that this coupling be interpreted as implying that certain signals are alternately multiplied by large and by small gain as they make successive trips around the feedback loop. A plausible conjecture is that a feedback system which is "fighting" itself in this way will indeed have poor feedback properties. Certainly, this conjecture seems consistent with the fact that such coupling implies both some disturbances *and* some sensor noise are amplified with gain greater than one. Equations (7.3.3c) and (7.3.4c) are obtained by decomposing the disturbance and sensor noise responses in terms of the outputs of the subsystems and have analogous interpretations.

It is interesting to note that these approximations suggest a method for assigning the invariant subspaces of S and T. Indeed, if sufficiently different levels of open loop gain are present, the directions of these subspaces are determined by those of the singular subspaces of L. This fact should prove useful in shaping the open loop transfer function to meet design specifications requiring that S and T have a desired set of invariant subspaces.

7.4. Generalizations of Gain Crossover Frequency for Each Subsystem

As discussed in Section 5.3, the relation between open loop gain and feedback properties is well-understood at frequencies for which gain is either large in all directions or small in all directions. Theorem 7.3.1 allows feedback properties to be approximated when open loop gain is both large and small in different directions at the same frequency. In this section we consider two additional cases: when gain is assumed to be either (a) large in some directions or (b) small in some directions, and, in each case, no assumption is placed upon gain in orthogonal directions. Equivalently, in terms of the subsystem decomposition (7.2.1), we are interested in what happens if one of the conditions (7.3.1)-(7.3.2) fails to hold. As motivation, consider a system with gain large in all directions over a low frequency range, but for which gain in some directions rolls off at a lower frequency than does gain in other directions (perhaps as a result of a design carried out for structured noise or uncertainty). The following theorem shows that there exists an analogue to scalar gain crossover frequency as the gain in the low gain subsystem in (7.2.1) rolls off. That is, there exists a frequency range in which knowledge of gain alone is insufficient to allow approximation of feedback properties; hence it is necessary to consider some sort of phase information.

Theorem 7.4.1: Assume that $\det[U_1^H V_1] \neq 0$ and that $\det[V_2^H U_2 + \Sigma_2] \neq 0$. Then, if

$$\underline{\sigma}[\Sigma_1] >> \max \begin{cases} \bar{\sigma}[(U_1^H V_1)^{-1}] \\ \bar{\sigma}\{(U_1^H V_1)^{-1}[I-(U_1^H V_2)\Sigma_2(V_2^H U_2+\Sigma_2)^{-1}(V_2^H U_1)]\} \\ \bar{\sigma}\{(U_1^H V_1)^{-1}(U_1^H V_2)\Sigma_2(V_2^H U_2+\Sigma_2)^{-1}(V_2^H U_2)\} \end{cases} \tag{7.4.1}$$

the sensitivity and complementary sensitivity functions satisfy $\bar{\sigma}[S-S_{app}] << 1$ and

$\bar{\sigma}[T-T_{app}] << 1$, where

$$S_{app} = U_2(V_2^H U_2 + \Sigma_2)^{-1} V_2^H \tag{7.4.2}$$

and

$$\begin{aligned} T_{app} &\equiv I - S_{app} \\ &= V_1(U_1^H V_1)^{-1} U_1^H + U_2(V_2^H U_2)^{-1}\Sigma_2(V_2^H U_2+\Sigma_2)^{-1} V_2^H \ . \end{aligned} \tag{7.4.3}$$

Proof: See [FrL86b]. ■

Theorem 7.4.1 shows that, at frequencies for which the conditions $\bar{\sigma}[\Sigma_2] << \underline{\sigma}[U_2^H V_2]$ and $\underline{\sigma}[\Sigma_2] >> \bar{\sigma}[U_2^H V_2]$ *both* fail to hold, knowledge of the open loop singular value "gains" does not provide sufficient information to allow feedback properties to be approximated. By analogy with the scalar case, some information about "phase" is needed. In general, this area requires further research. In certain cases, however, more may be said immediately. First, note that the condition for the low gain subsystem to be rolled off is that $\bar{\sigma}[\Sigma_2] << \underline{\sigma}[V_2^H U_2]$. Since $\underline{\sigma}[V_2^H U_2] \leq 1$, this is a more stringent requirement than the scalar system analogue which only requires $|L| << 1$. Indeed, if $\underline{\sigma}[V_2^H U_2] << 1$, then the gain $\bar{\sigma}[\Sigma_2]$ will also have to be much less than one before the second subsystem can be considered "rolled off". This might seem to limit the applicability of the above results by extending the frequency range over which phase information is necessary and Theorem 1 does not hold. However, if $\underline{\sigma}[V_2^H U_2] << 1$ *and* $\bar{\sigma}[\Sigma_2] << 1$, then (7.4.2) shows that $\bar{\sigma}[S_{app}] >> 1$, *regardless* of any consideration of phase.

Suppose that the low gain subsystem has rank one. Then the subspaces $\mathbf{V}_2$ and $\mathbf{U}_2$ are each one-dimensional and, in fact, are equal to the left and right singular subspaces associated with the smallest singular value of L. Since, by convention, this singular value is denoted $\underline{\sigma}[L] \triangleq \sigma_n[L]$, we shall denote the corresponding singular vectors and subspaces by $\underline{v}$, $\underline{u}$, $\underline{\mathbf{V}}$, and $\underline{U}$, respectively. Hence the approximation to the sensitivity function becomes $S_{app} = \underline{u}(\underline{v}^H \underline{u} + \underline{\sigma})^{-1}\underline{v}^H$. The angle between the subspaces $\underline{V}$ and $\underline{U}$ is equal to the angle (6.5.2) between the left and right singular subspaces V_2 and U_2, and will be denoted $\underline{\phi} \triangleq arccos\, |\underline{u}^H \underline{v}|$. We shall also denote the phase difference (6.5.1) between the singular vectors $\underline{v}$ and $\underline{u}$ by $\Delta\underline{\theta} \triangleq \measuredangle \underline{u}^H \underline{v}$.

Suppose that we give the open loop system an input in the direction of the right singular subspace $\underline{U}$, and consider the resulting steady-state output, which must lie in the left singular subspace $\underline{V}$. We saw in Chapter 6 that the angle $\underline{\phi}$ between these two subspaces determines the extent to which these signals can interfere when added together. The phase difference $\Delta\underline{\theta}$ between the input and output signals determines whether the signals interfere more or less constructively or destructively. As one might expect, the parameters $\underline{\sigma}$, $\underline{\phi}$, and $\Delta\underline{\theta}$ may be used to obtain a lower bound (6.5.3) on the size of the sensitivity matrix of the feedback system. Clearly, additional information is needed to evaluate sensitivity properties in other directions. Our assumption in this section that $\underline{\sigma}$ is much smaller than the other singular values allows some additional information to be obtained:

Corollary 7.4.2: Suppose the subspaces U_2 and V_2 in Theorem 7.4.1 have dimension one. Then

$$S_{app} = e^{j\Delta\underline{\theta}}\underline{u}(\cos\underline{\phi} + \underline{\sigma} e^{j\Delta\underline{\theta}})^{-1}\underline{v}^H \quad , \qquad (7.4.4)$$

where $\underline{\sigma} \triangleq \underline{\sigma}[L]$, $\Delta\underline{\theta} = \measuredangle\, \underline{u}^H \underline{v}$, and $\underline{\phi} = arccos\, |\underline{u}^H \underline{v}|$.

■

Note Corollary 7.4.2 implies that $\overline{\sigma}[S] \approx \dfrac{1}{|\cos\underline{\phi} + \underline{\sigma} e^{j\Delta\underline{\theta}}|}$. Hence, to maintain good feedback properties at frequencies for which $\underline{\sigma} \approx \cos\underline{\phi}$, it is necessary and sufficient to require that $\Delta\underline{\theta}$ be bounded away from ±180°. This result is appealing in that it shows how the intuitive notion of phase difference defined in Chapter 6 can be directly related to feedback properties of the system.

When approximation (7.4.2) holds, it follows that k of the singular values of S are very small, while the remaining $n-k$ singular values are approximated by

$$\sigma_i[S] = \sigma_i[(V_2^H U_2 + \Sigma_2)^{-1}] \quad ; \qquad i = 1, \ldots, n-k \ . \tag{7.4.5}$$

Approximating the singular values of T is more difficult since (7.4.3) is not in the form of a singular value decomposition. Some useful information may still be obtained, however. From the identity $T + S \equiv I$ and Theorem 6.6 of [Ste73a], it follows that $|\sigma_i[S] - \sigma_i[T]| \leq 1$. In particular, $\sigma_i[S]$ is large if and only if $\sigma_i[T]$ is large.

An alternate approximation may be obtained from using the identities $V_1 V_1^H + V_2 V_2^H \equiv I$ and $T_{app} = I - S_{app}$; this yields $T_{app} = V_1 V_1^H + [-V_1(V_1^H U_2) + V_2 \Sigma_2]\ (V_2^H U_2 + \Sigma_2)^{-1} V_2^H$. Since the nonzero singular values of $V_1 V_1^H$ are equal to one, it follows that $\sigma_i[T_{app}] >> 1$ if and only if $\sigma_i\{[-V_1(V_1^H U_2) + V_2\Sigma_2](V_2^H U_2 + \Sigma_2)^{-1} V_2^H\} >> 1$. In this case

$$\sigma_i[T] = \sqrt{\sigma_i^2[(V_1^H U_2)(V_2^H U_2 + \Sigma_2)^{-1}] + \sigma_i^2[\Sigma_2(V_2^H U_2 + \Sigma_2)^{-1}]} \tag{7.4.6}$$

This approximation is useful when it is desired to detect large values of $\sigma_i[T]$, since the approximation is valid precisely in this case.

Finally, note that the magnitude of $\overline{\sigma}[(V_2^H U_2 + \Sigma_2)^{-1}]$ is related, via (7.4.1), to the level of gain required in the high gain subsystem to insure that the approximations hold. The bounds (7.4.1) may appear too complicated to be of use. We suggest, however, that one not be too pedantic about conditions (7.4.1). Rather, Theorem 7.4.1 should be regarded as giving us a *rule of thumb* concerning what happens to feedback properties when gain is allowed to become large in only *some* directions or loops and not in others.

The following theorem is the counterpart of Theorem 7.4.1 applicable when the high gain subsystem rolls off at a frequency beyond the crossover region for the lower gain subsystem.

Theorem 7.4.3: Assume that $\det[V_2^H U_2] \neq 0$ and that $\det[I + U_1^H V_1 \Sigma_1] \neq 0$. Then, if

$$\bar{\sigma}[\Sigma_2] << \min \begin{cases} \underline{\sigma}[V_2^H U_2] \\ \underline{\sigma}\{[I-(V_2^H U_1)(I+U_1^H V_1 \Sigma_1)^{-1}(U_1^H V_2)]^{-1}(V_2^H U_2)\} \\ 1/\bar{\sigma}\{(V_2^H U_2)^{-1}(V_2^H U_1)(I+U_1^H V_1 \Sigma_1)^{-1}(U_1^H V_1)\} \end{cases} \tag{7.4.7}$$

the sensitivity and complementary sensitivity functions satisfy $\bar{\sigma}[S - S_{app}] << 1$ and $\bar{\sigma}[T - T_{app}] << 1$, where

$$\begin{aligned} S_{app} &\equiv I - T_{app} \\ &= U_2^H (V_2^H U_2)^{-1} V_2^H + V_1 (U_1^H V_1)^{-1} (I + U_1^H V_1 \Sigma_1)^{-1} U_1^H \end{aligned} \tag{7.4.8}$$

$$T_{app} = V_1 \Sigma_1 (I + U_1^H V_1 \Sigma_1)^{-1} U_1^H \tag{7.4.9}$$

Proof: See [FrL86b].

■

Comments similar to those following Theorem 7.4.1 are applicable to Theorem 7.4.3 as well. Note that gain crossover for the high gain subsystem occurs when the conditions $\underline{\sigma}[\Sigma_1] >> \bar{\sigma}[(U_1^H V_1)^{-1}]$ and $\bar{\sigma}[\Sigma_1] << \underline{\sigma}[(U_1^H V_1)^{-1}]$ both fail to hold. Since $\bar{\sigma}[(U_1^H V_1)^{-1}] \geq 1$, gain crossover frequency occurs for a larger value of gain than for scalar systems. Hence phase information may have to be considered for values of gain greater than one. In particular, note that if $\bar{\sigma}[(U_1^H V_1)^{-1}] >> 1$, the gain $\underline{\sigma}[\Sigma_1]$ will also have to be much greater than one for condition (7.3.1) in Theorem 7.3.1 to hold. However, if both $\bar{\sigma}[(U_1^H V_1)^{-1}] >> 1$ *and* $\underline{\sigma}[\Sigma_1] >> 1$, then (7.4.9) may be manipulated to show that $\bar{\sigma}[T_{app}] >> 1$, *regardless* of phase considerations.

In the case that the high gain subsystem has rank equal to one, a result analogous to Corollary 7.4.2 may be obtained. In this case, the feedback properties of the system are determined by the largest singular value of L and the associated left and right singular vectors and subspaces, which we denote $\bar{\sigma}[L]$, $\bar{v}$, $\bar{u}$, $\bar{\mathbf{V}}$, and $\bar{\mathbf{U}}$, respectively. The following corollary relates the complementary sensitivity function to the angle $\bar{\phi} \triangleq arccos\,|\bar{u}^H \bar{v}|$ between the singular subspaces and the phase difference $\overline{\Delta\theta} = \measuredangle\, \bar{u}^H \bar{v}$ between the singular vectors.

Corollary 7.4.4: Suppose the subspaces $\mathbf{U}_1$ and $\mathbf{V}_1$ in Theorem 7.4.3 have dimension one. Then

$$T_{app} = \bar{\sigma} \cdot \bar{v}(1+\bar{\sigma}e^{j\overline{\Delta\theta}}\cos\bar{\phi})^{-1}\bar{u} \quad , \tag{7.4.10}$$

where $\bar{\sigma} \triangleq \bar{\sigma}[L]$.

■

Note Corollary 5 implies that $\bar{\sigma}[T] = \bar{\sigma}/\,|1+\bar{\sigma}e^{j\overline{\Delta\theta}}\cos\bar{\phi}|$. Hence, at frequencies for which $\bar{\sigma} = 1/\cos\bar{\phi}$, the phase difference $\overline{\Delta\theta}$ plays a role in determining feedback properties analogous to that played by scalar phase near gain crossover.

Finally, note for later reference that the analogues to (7.4.5) and (7.4.6) are

$$\sigma_i[T] = [\Sigma_1(I+U_1^HV_1\Sigma_1)^{-1}] \qquad i = 1, \ldots, k \tag{7.4.11}$$

and

$$\sigma_i[S] = \sqrt{\sigma_i^2[(I+U_1^HV_1\Sigma_1)^{-1}] + \sigma_i^2[(U_2^HV_1)\Sigma_1(I+U_1^HV_1\Sigma_1)^{-1}]} \ . \tag{7.4.12}$$

We close this section with a brief discussion of what happens when $\det[U_1^HV_1] = 0$ in Theorem 1. Since $\det[U_1^HV_1] = 0$ if and only if $\det[U_2^HV_2] = 0$, it is clear that neither condition (7.3.1) nor (7.3.2) can be satisfied. Analysis of this case using pseudoinverses appears to result in excessively complicated formulas. The type of behavior which can take place may be seen from considering the two-input two-output case.

Theorem 6: Assume $L(s) \in \mathbb{C}^{2\times 2}$ and that $u_i^H v_i = 0$, $i = 1,2$. If $\sigma_1[L] >> 1 >> \sigma_2[L]$, then $\bar{\sigma}[S - S_{app}] << 1$, where

$$\bar{\sigma}[S_{app}] \geq \frac{1}{\frac{1}{\sigma_1[L]} + \sigma_2[L]} \quad . \tag{7.4.13}$$

Proof: See [FrL86b]. ■

The bound (7.4.13) shows that $\bar{\sigma}[S] >> 1$ if both high and low loop gains are used in a system where the outputs of each subsystem are fed *completely* into the inputs of the other subsystem. That excessively poor feedback properties should result in this case is to be expected from the thought experiment.

7.5. A Practical Example

The purpose of this section is to use the results of this chapter to analyze an actual physical design. The motivation for doing so is two-fold. First, the analysis will allow us to verify that the approximations derived above are actually useful for practical systems. Second, the analysis will show how a designer might use the approximations to gain insight into design tradeoffs and limitations.

The example considered is described in[ALP83]. The system has two inputs and outputs and consists of a laser beam whose direction is controlled by two mirrors. Physical considerations dictated that the gain in one loop of the system be rolled off at about a decade below that in the other loop. Over the intermediate frequency range, between crossover for the two loops, a strong degree of loop coupling was introduced in order to reduce torque requirements on the mirrors. A design meeting these constraints was completed and an analysis of feedback properties was carried out using the singular value decomposition of the system sensitivity matrix.

The conclusion reached was that the system had unacceptably large disturbance response properties as well as unacceptably small stability margins. Since analyses based upon breaking the loops one at a time indicated good feedback properties, it was concluded that the difficulty was due to the coupling.

It must be emphasized that the analysis in [ALP83] was performed *a posteriori* and the fact that coupling led to a poor design was discovered only after the design was completed. In the remainder of this section, the results of this chapter will be used to describe how an analysis of feedback properties may be carried out *a priori*, and the tradeoff between the level of loop coupling and feedback properties may be seen without completing the details of the design. Indeed, this example is just a special case of our general result that highly coupled systems with a large spread in gains can have large disturbance and noise response. The following analysis is a summary of a more detailed discussion in[Fre85]. Although this example only has two loops, much of the analysis can be extended directly to larger systems decomposed as in (7.2.1).

To illustrate the generality of our analysis, we shall only work with two general features of the system. Denote $L = [L_1 | L_2] = \begin{bmatrix} L_{11} L_{12} \\ L_{21} L_{22} \end{bmatrix}$. The relevant feature of this example is that there exists a frequency range over which (a) $||L_1|| >> ||L_2||$, *and* (b) $|L_{21}| > |L_{11}|$.

First, note that the minimax characterization of singular values implies that

$$\sigma_1[L] \geq ||L_1|| \text{ and } ||L_2|| \geq \sigma_2[L] \ . \tag{7.5.1}$$

Hence (a) implies that $\sigma_1[L] >> \sigma_2[L]$ at frequencies of interest. Furthermore, denoting the first right singular vector as $u_1 = [u_{11}\ u_{21}]^T$, it follows that

$$\sigma_1[L] \leq |u_{11}| \cdot ||L_1|| + ||L_2|| \tag{7.5.2}$$

Together, (7.5.1) and (7.5.2) imply

$$|u_{11}| \geq 1 - \frac{||L_2||}{||L_1||} \quad ;$$

hence, (a) implies that $|u_{11}| \approx 1$. In fact, it is easy to see that $\sigma_1[L] \approx ||L_1||$, and that the right singular vectors may be well approximated by the standard basis vectors: $u_i \approx e_i$.

Next, since $Lu_1/\sigma_1 = v_1 = (u_{11}L_1+u_{21}L_2)/\sigma_1$, it follows from (a) that $||v_1-u_{11}L_1/\sigma_1|| = ||u_{21}L_2||/\sigma_1 << 1$. Hence $v_1 \approx L_1/\sigma_1$ and $|u_1^H v_1| \approx |u_1^H L_1|/\sigma_1 \approx \frac{1}{\sqrt{1+|L_{21}/L_{11}|^2}}$. Noting that $u_1 = \overline{u}$ and $v_1 = \overline{v}$, it follows that the angle $\overline{\phi}$ defined in Section 7.4 is approximately

$$\overline{\phi} \approx arccos\left[\frac{1}{\sqrt{1+|L_{21}/L_{11}|^2}}\right] \quad . \tag{7.5.3}$$

Equivalently,

$$\overline{\phi} \approx arctan\,|L_{21}/L_{11}| \quad .$$

Finally, recall that properties of the principal angles discussed in Section 7.2 imply $\underline{\phi} = \overline{\phi}$; i.e., the angles between the two pairs of singular subspaces are identical, and will be denoted ϕ.

The preceding remarks show that at frequencies between crossover for the two subsystems (crossover being defined by $\sigma_1 \approx 1/\cos\phi$ and $\sigma_2 \approx \cos\phi$) *and* for which (a) holds, Theorem 7.3.1 implies

$$\begin{aligned} \overline{\sigma}[S] &\approx 1/\cos\phi \\ &\approx \frac{1}{\sqrt{1+|L_{21}/L_{11}|^2}} \quad . \end{aligned} \tag{7.5.4}$$

Since $|L_{21}/L_{11}|$ is a measure of coupling, (7.5.4) shows that there exists a design tradeoff: as coupling between loops increases, sensitivity properties become worse.

Furthermore, from (b) we know that the coupling is fairly strong; hence, we can anticipate a problem, either in the mid-frequency range or near one of the crossover frequencies. More cannot be said without considering the actual data for this example.

The open loop transfer function for this system is

$$L(s) = \frac{(2600)(2\pi)^2}{s^2} \begin{bmatrix} 1 & 0 \\ \dfrac{(-1.825s + 19.5\pi)}{(s + 300\pi)} & 2.52 \end{bmatrix} \begin{bmatrix} G_\alpha & 0 \\ 0 & G_y \end{bmatrix}$$

$$G_\alpha = (172.8)\left[\frac{s + 1068.7}{s + 7480.7}\right]\left[\frac{s + 188.5}{s}\right] \qquad (7.5.5)$$

$$G_y = (1.782)\left[\frac{s + 166.2}{s + 1163.7}\right]\left[\frac{s + 62.8}{s}\right]$$

Bode plots of the $L_{ij}(s)$ reveal that gain crossover for L_{11} is $\approx$ 2300 *rad/sec*, for L_{22} is $\approx$ 400 *rad/sec*, and for L_{21} is $\approx$ 4000 *rad/sec*, clearly indicating both a spread in loop gains as well as strong coupling. The effect that the spread in loop gains has on the singular values is seen from Figure 7.5.1. Figure 7.5.2 shows that coupling is indeed strong; ϕ approaches 60°. Approximations to $\underline{\sigma}[I+L] = 1/\overline{\sigma}[S]$ are shown in Figure 7.5.3. Recall $\underline{\sigma}[I+L]$ is a measure of stability margin and is plotted to be consistent with[ALP83]. The first approximation shown is (7.4.2) and, as expected, holds fairly well at low to mid frequencies. The second approximation is (7.4.12), and holds at mid to high frequencies. The approximations predict the actual value of $\underline{\sigma}[I+L]$ much more accurately than estimates of stability margin obtained by breaking the loops one at a time (Figure 7.5.4). More significantly, they give *a priori* insight into how the loop coupling purposely introduced into the design deteriorates feedback properties.

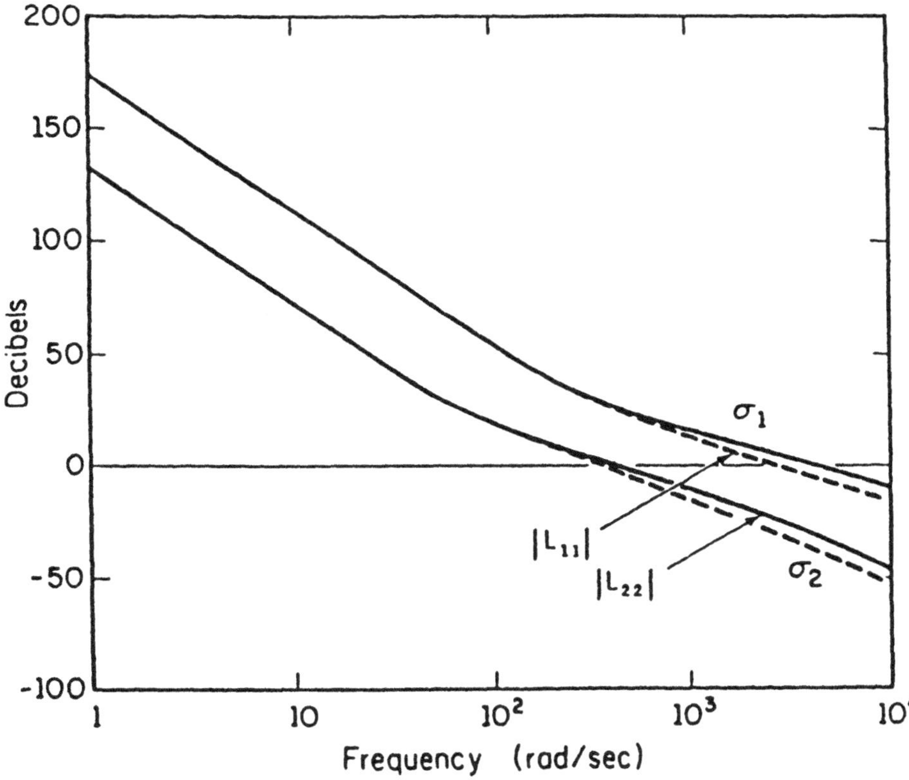

Figure 7.5.1: Open Loop Singular Values

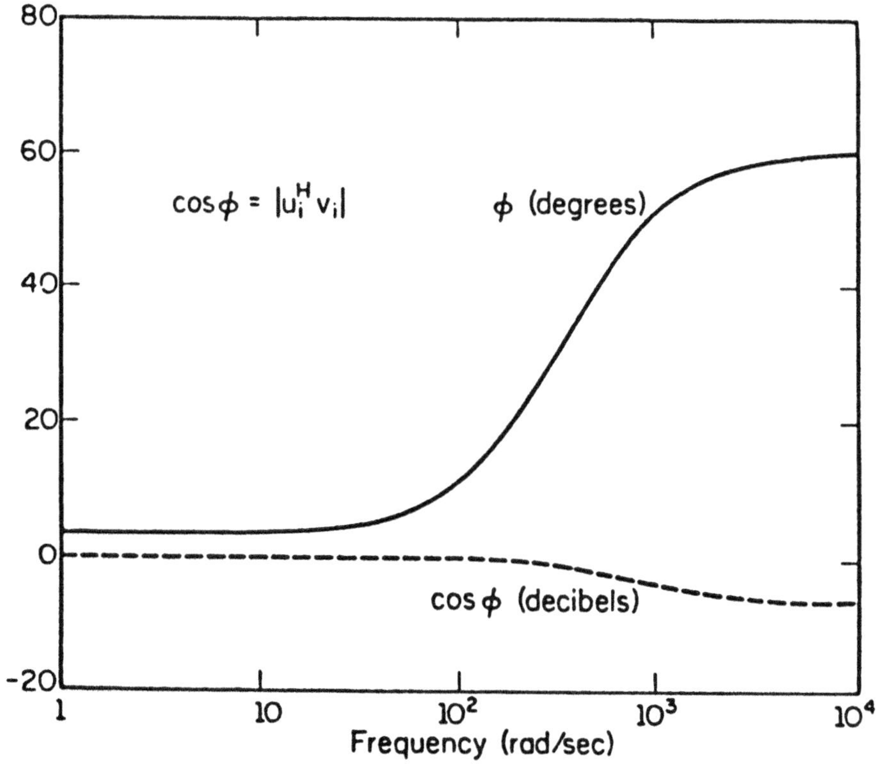

Figure 7.5.2: Angle Between Singular Subspaces

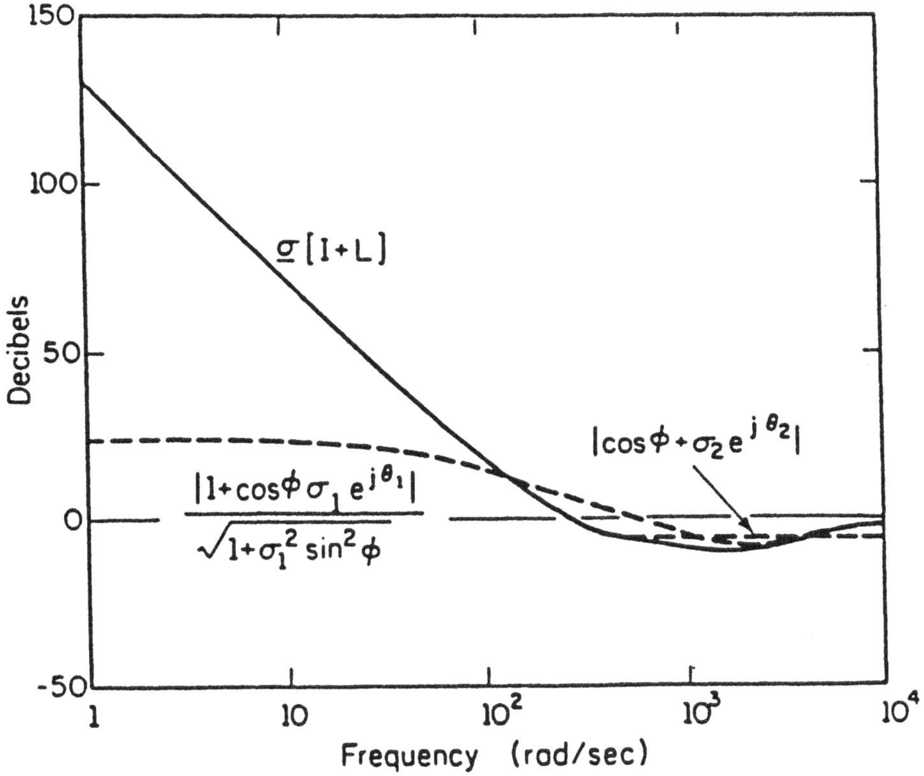

Figure 7.5.3: The Stability Margin $\underline{\sigma}[I+L]$ and Two Approximations

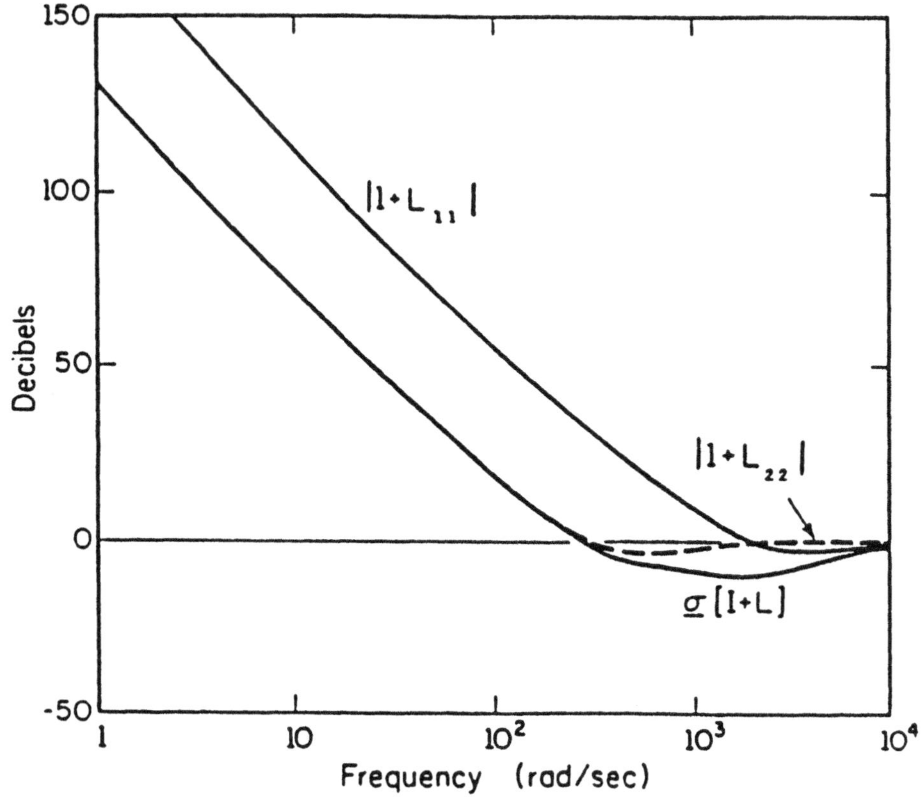

Figure 7.5.4: Single Loop Stability Margins vs. Singular Value Stability Margin

7.6. Summary

In this chapter we have discussed feedback properties of systems whose open loop transfer functions may be decomposed into high and low gain "subsystems" at a frequency of interest. We have shown that feedback properties are a function of the coupling between such subsystems and that frequency ranges analogous to gain crossover frequency exist for each subsystem. These results provide insight into *one* of the difficulties posed by plants with poorly conditioned transfer function matrices. Since poorly conditioned plants have, by definition, much higher levels of gain in some directions than in others, it follows that such a system can potentially have poor feedback properties unless care is taken that the high and low gain portions of the plant are sufficiently decoupled. If the coupling between high and low gain directions of the plant is sufficiently strong, then it will be necessary to achieve the decoupling through an appropriate compensation scheme.

The results of this chapter should prove useful in feedback design when uncertainty, noise and/or disturbances exhibit structure. To illustrate, let us return to the structured sensor noise problem discussed in Section 5.4. We saw that the structure exhibited in Figure 5.4.1 could be translated into the design specification (5.4.4). Over the frequency range (ω_2, ω_3) the sensitivity and complementary sensitivity functions are each required to have small gain, but in different directions. The rules of thumb (5.3.10-11) relating open loop gain to feedback properties yield no insight into how this specification might be achieved. Intuition suggests that we should require open loop gain to be small in the direction of the structured sensor noise and large in the orthogonal direction. Using the projection operators introduced in Section 5.4, we can state this requirement mathematically as

$$\overline{\sigma}[L(j\omega)P_s] \ll 1 \quad , \quad \omega \in (\omega_2, \omega_3) \tag{7.6.1a}$$

and

$$\underline{\sigma}[L(j\omega)P_s^{\perp}] \gg 1 \quad , \quad \omega \in (\omega_2, \omega_3) \quad . \tag{7.6.1b}$$

This specification may be restated in terms of the decomposition (7.2.1) into high and low gain subsystems. We require that $\underline{\sigma}[\Sigma_1] \gg 1$, $\overline{\sigma}[\Sigma_2] \ll 1$, and $\mathbf{U}_2 = \mathbf{N}_s$ (which implies $\mathbf{U}_1 = \mathbf{N}_s^{\perp}$).[23] The first two conditions insure that the open loop system has both high and low gains and the third insures that the structured noise enters the system in the direction of low gain. This much of the design requirement is suggested by intuition gleaned from experience with scalar systems.

Approximation (7.3.4a) shows that if the singular values and subspaces of L satisfy these requirements, then the structured noise will indeed be rejected. Furthermore, (7.3.3b) shows that disturbances in the subspace $\mathbf{N}_s$ are unaffected by the presence of feedback and appear directly in the system output. Of course, the identity $T+S \equiv I$ shows that this is to be expected. It is important to note that noise and disturbances in the subspace $\mathbf{N}_s^{\perp}$ are amplified by amounts $\overline{\sigma}[(V_2^H U_2)^{-1}(V_2^H U_1)] \approx \tan\overline{\alpha}$ and $\overline{\sigma}[T] \approx 1/\cos\overline{\alpha}$, respectively, where $\overline{\alpha}$ is the largest principal angle between the subspaces $\mathbf{U}_i$ and $\mathbf{V}_i$. Hence, to insure that use of high gain is successful in rejecting disturbances orthogonal to the structured noise (so that (5.4.4c) is satisfied), we must require that the left and right singular subspaces be sufficiently well-aligned, so that $\overline{\alpha}$ is small. This strategy also prevents amplification of the unstructured noise present in directions orthogonal to the structured noise. To summarize, recall our intuition which suggested that we could (a) reject the structured component of the sensor noise by requiring small loop gain in the direction of that noise and (b) reject disturbances in orthogonal directions by requiring large loop gain in those directions. We have shown that this intuition is verified *provided* we impose the additional requirement that the resulting high and low gain "subsystems" be sufficiently decoupled.

---

[23] Recall that $\mathbf{N}_s$ is the k-dimensional subspace containing the structured noise.

For later reference, it is important to note that the high gain requirement (7.6.1b) will *always* tend to imply that the sensitivity function has small gain on an $(n-k)$-dimensional subspace:

$$\sigma_i[S] \ll 1 \quad , \quad i = k+1, \ldots, n \quad . \tag{7.6.2}$$

Furthermore, this statement holds true even when $\overline{\sigma}[S] \gg 1$ due to coupling between high and low gain loops. Indeed, if (7.6.1b) holds, then

$$\overline{\sigma}[P_s^{\perp} S] \ll 1 \, , \quad \omega \in (\omega_2, \omega_3) \quad . \tag{7.6.3}$$

Hence, rather than requiring that (5.4.4c) hold, we could require that disturbances in a subspace *other* than $\mathbf{N}_s^{\perp}$ be rejected; this may be done by imposing an appropriate restriction on the left singular subspaces of L. Of course, if the resulting high and low gain "subsystems" have input and output directions that are sufficiently misaligned, then both S and T will have large gains in other directions.

We also discussed, in Section 5.4, the design specification (5.4.10) arising from structured uncertainty. Again, approximation (7.3.4) shows that appropriate shaping of the open loop transfer function can yield a system satisfying this specification. Denote the columnspaces of X and Y by $\mathbf{X}$ and $\mathbf{Y}$, respectively. Then requiring that L have low gain in the input direction $\mathbf{U}_2 \approx \mathbf{Y}$ and/or in the output direction $\mathbf{V}_2 \approx \mathbf{X}$ and large gain in the orthogonal direction implies that $\overline{\sigma}[X^T T Y] \ll 1$ as desired. Equations (7.3.3-4) also show, however, that if $\mathbf{U}_i$ and $\mathbf{V}_i$ are not sufficiently well-aligned, then feedback properties will be poor in orthogonal directions. Consequently, just as in the structured sensor noise problem, care must be taken that high and low gain subsystems be sufficiently decoupled.

One significant shortcoming of the results in this chapter is that they deal only with system properties at a *given* frequency. The difficulty with this is that system properties in different frequency ranges are not independent of one another, as is shown by the Bode integral relations [Bod45] and Poisson integral formulas[FrL85a]. This dependence implies that design specifications imposed assuming independence may not be achievable and system properties in different frequency ranges may have to be traded off against one another. In the context of this chapter, the most relevant of these relations is the Bode gain-phase relation, whose implications for design we discussed in Chapter 2. At least in some cases, our results may be used to state design specifications in terms of singular value "gains" and measures of "phase difference" between the associated pair of singular vectors. Hence, it is only natural to ask whether these quantities also must satisfy a gain-phase relationship. If so, then this dependence will have to be taken into account in developing design specifications. Since the generalization of the gain-phase analysis discussed in Section 5.3 does not apply to systems with different levels of gain in different directions, we are therefore motivated to seek additional generalizations. In fact, we shall devote the remainder of the monograph to developing a mathematical framework within which to study such problems.

CHAPTER 8

SINGULAR VALUES AND ANALYTIC FUNCTION THEORY

8.1. Introduction and Motivation

Throughout this monograph we have studied the existence of various design tradeoffs limiting the ability of a linear feedback system to achieve design goals. These tradeoffs divide naturally into two types. "Algebraic" tradeoffs, so-called because the underlying equations are algebraic, exist between feedback system properties at a single frequency. We discussed these tradeoffs in the context of multivariable systems in Chapters 5 and 7.

The other type of tradeoffs we termed "analytic," since the equations governing them are derived using complex variable theory. Analytic tradeoffs are considerably more subtle than the algebraic ones, as they take place between system properties in different frequency ranges. For scalar systems these tradeoffs are quantified by the Bode and Poisson integrals discussed in Chapters 2, 3, and 4. Some simple extensions to multivariable systems were discussed in Chapter 5. One difficulty with these extensions is that they give useful information only in special cases. It would be much preferable if we could express multivariable design tradeoffs using parameters more readily related to feedback properties.

Toward this end, recall the singular value decomposition discussed in Chapters 5-7. We saw that, in many cases, important feedback properties can be directly related to the singular values and vectors of various transfer function matrices. Indeed, we can even pose design specifications in terms of the singular value decomposition. Once such a specification is posed, however, it is necessary to investigate whether it is achievable and, if not, what design tradeoffs must be performed to arrive at an achievable specification. In particular, we would like to discover whether tradeoffs analogous to those imposed by the Bode and Poisson integrals exist. If so, we would like to quantify them using the singular values and vectors upon which design specifications are stated. The remainder of this monograph is devoted to developing the mathematical tools with which to study such questions. Although we restrict our attention to a multivariable version of the tradeoff imposed by the scalar gain-phase relation, the mathematical framework we shall develop should also prove useful in studying multivariable versions of the Bode sensitivity integral and the Poisson integral relations. Hence, even if the tradeoff imposed by the gain-phase relation is equivalent to that imposed by the sensitivity integral (in the sense discussed in Chapter 4), additional analysis methods are still needed to study multivariable systems.

The Bode gain-phase relation has at least two useful interpretations. One is that not all desirable design specifications are achievable. The other is that partial knowledge of system properties implicitly contains knowledge of the remaining properties. In Section 2.5 we discussed how these facts limit the rate at which open loop gain may be rolled off near crossover if a robustly stable feedback system is to be achieved. Motivated by the discussion of structured multivariable design specifications in Section 5.4, it is reasonable to ask: What happens when the gain in one loop of a multivariable system rolls off rapidly near crossover if the gain in other loops remains large? In the scalar case, if loop gain decreases monotonically at $\approx 40db/decade$ in the vicinity of gain crossover, the phase asymptotically approaches $-180°$ and the sensitivity function has a large peak at crossover. We would like to know if it possible to monotonically decrease the gain in one loop of a multivariable system at $40db/decade$ and either prevent such a peak from occurring or to reduce its size significantly. We shall see, eventually, that the answer to this question is yes. First, however, we must investigate some much simpler questions, such as: Given that we know the behavior of one singular value of a matrix transfer function along the $j\omega$-axis, what, if any, additional knowledge does this imply about properties of the matrix?

For scalar systems, the answer to such questions lies in use of analytic function theory. Loosely speaking, this theory shows that the two real parameters needed to describe the value of a scalar transfer function (real and imaginary parts, or gain and phase) yield only one degree of freedom to be manipulated in design. This fact may be seen from the Cauchy-Riemann equations ([LeR70, Con78]) or from the Hilbert transform, as applied to network design by Lee and Wiener[Lee32]. It was not until Bode's work[Bod45], however, that the dependence between gain and phase was displayed in a form yielding the most useful insights for feedback design.

Consider next an $n \times n$ transfer function matrix written in standard coordinates. Since each element of this matrix is an analytic function, it follows that, loosely speaking, the $2n^2$ real parameters needed to describe the value of the matrix can yield only n^2 degrees of freedom to be manipulated in design. This statement can be made precise by applying the scalar integral relations to the gain and phase of each element of the matrix. Unfortunately, the standard coordinates of a matrix transfer function give useful information about feedback properties only in special cases. Indeed, this is one of our motivations for developing alternate sets of coordinates. Toward this end, we have seen in Chapters 5 and 7 that the singular values and vectors of a matrix transfer function do yield useful information about feedback properties. It is straightforward to verify that $2n^2$ real parameters are needed to describe the singular value decomposition of a matrix $M \in \mathbb{C}^{n \times n}$. Clearly, these parameters must also yield only n^2 degrees of freedom useful for design. Hence, there must exist limitations upon the ability of a linear system to achieve specifications imposed upon the singular values and vectors of its transfer function.

From a purely mathematical standpoint, the structure imposed upon a matrix transfer function by the property of complex analyticity is already completely understood. Indeed, in standard coordinates, this structure may be described using the Cauchy-Riemann equations, Cauchy Integral Theorem, or the scalar gain-phase relations applied to each element of the matrix. From a practical standpoint, it remains to determine how this structure manifests itself after transformation to a set of coordinates useful in analysis and design. Since we adopt the point of view that parameterizations of the singular value decomposition are useful for these purposes, we shall therefore proceed to study how the familiar results of scalar complex variable theory appear when transformed from standard coordinates to coordinates describing the singular values and vectors.

The remainder of this chapter is organized as follows. In Section 8.2 a review of elementary facts from complex variable theory is presented. The analogy between singular values and gain suggests that a singular value might be the *magnitude* of an analytic function. In Section 8.3, this conjecture is studied using facts from Section 8.2. An interpretation and summary of the result is contained in Section 8.4.

8.2. Review of Analytic Function Theory

The Bode gain-phase relation expresses the interdependence of scalar gain and phase in the form most useful in applications to feedback theory. In attempting to generalize this relation to multivariable systems, it appears more tractable to first work with this interdependence as expressed by the Cauchy-Riemann equations and properties of harmonic functions. To do this, we need some elementary results from complex variable theory ([LeR70, Con78]).

Definition 8.2.1[Con78]: Let D be an open subset of $\mathbb{C}$, the complex s-plane. Then a function $f : D \rightarrow \mathbb{C}$ is *analytic* if f is continuously differentiable on D; i.e., if the derivative of f with respect to s exists and is continuous at all points of D.

■

Definition 8.2.2[Con78]: A *region* is an open connected subset of the complex plane.

■

Theorem 8.2.3[Con78]: Let g and h be real-valued functions defined on a region D of the complex s-plane, with $s = x + jy$. Suppose that g and h have continuous first partial derivatives. Then the function $f : D \rightarrow \mathbb{C}$ defined by $f(s) \triangleq g(s) + jh(s)$ is analytic if and only if g and h satisfy the **Cauchy-Riemann equations:**

$$\begin{aligned} \frac{\partial g}{\partial x} &= \frac{\partial h}{\partial y} \\ \frac{\partial g}{\partial y} &= -\frac{\partial h}{\partial x} \end{aligned} \tag{8.2.1}$$

Equivalently, $\frac{\partial f}{\partial x} = \frac{1}{j}\frac{\partial f}{\partial y}$.

■

Since we wish to work with the log magnitude and phase of a transfer function, we need to be able to define the logarithm of an analytic function, when this is possible.

Definition 8.2.4[Con78]: An open set D is *simply connected* if D is connected and every closed curve in D is homotopic to zero.

■

Theorem 8.2.5[Con78]: Let D be an open connected subset of $\mathbb{C}$. Then the following statements are equivalent

(i) D is simply connected

(ii) For any function f which is analytic and nonzero on D there exists a function $\log f$, also analytic on D, such that $f(s) = \exp[\log f(s)]$.

■

In fact, there are infinitely many functions satisfying (ii), each of which is termed a *branch* of the logarithm. Given such a branch, we denote its real and imaginary parts by

$$\log f = \log|f| + j \measuredangle f \quad , \tag{8.2.2}$$

The real-valued functions $\log|f|$ and $\measuredangle f$ are termed the log magnitude and phase of f, respectively, and satisfy

$$\begin{aligned} \frac{\partial \log|f|}{\partial x} &= \frac{\partial \measuredangle f}{\partial y} \\ \frac{\partial \log|f|}{\partial y} &= -\frac{\partial \measuredangle f}{\partial x} \end{aligned} \tag{8.2.3}$$

Suppose $f(s)$ is analytic and nonzero on a simply connected region D. Then equations (8.2.3) may be used to show that knowledge of the gain of $f(s)$ suffices to determine the phase of $f(s)$ to within a constant. Indeed, given a branch of $\log f$ and the value of $\measuredangle f$ at some point $s_0 \in D$, the value of $\measuredangle f$ at any other point $s \in D$ can be computed by a line integral. To see this, let $\gamma: [0,1] \rightarrow D$ be a piecewise smooth path with $\gamma(0) = s_0$ and $\gamma(1) = s$. Then $\measuredangle f(s)$ can be found by integrating the differential form[24]

$$\begin{aligned} d\measuredangle f &= \frac{\partial \measuredangle f}{\partial x} dx + \frac{\partial \measuredangle f}{\partial y} dy \\ &= \frac{-\partial \log|f|}{\partial y} dx + \frac{\partial \log|f|}{\partial x} dy \\ &= * \, d \log|f| \end{aligned} \tag{8.2.4}$$

---

[24]The properties of differential forms used in this monograph are reviewed in Appendix B.

along γ. Thus

$$\begin{aligned}
\measuredangle f(s) &= \measuredangle f(s_0) + \int_\gamma d\measuredangle f \\
&= \measuredangle f(s_0) + \int_0^1 \left[\frac{\partial \measuredangle f}{\partial x}\frac{dx}{dt} + \frac{\partial \measuredangle f}{\partial y}\frac{dy}{dt} \right] dt \\
&= \measuredangle f(s_0) + \int_0^1 \left[-\frac{\partial \log|f|}{\partial y}\frac{dx}{dt} + \frac{\partial \log|f|}{\partial x}\frac{dy}{dt} \right] dt \ .
\end{aligned} \tag{8.2.5}$$

From (8.2.5) it is clear that the log magnitude of f determines the phase of f to within a constant, although it is difficult to obtain any insight into the qualitative nature of this relationship.

The fact that $\measuredangle f(s)$ is well-defined in a simply connected region once its value at one point is given follows since the line integral in (8.2.5) is independent of the path γ. Path independence in a simply connected region is guaranteed if the differential form (8.2.4) is closed; i.e., if it satisfies the condition (Appendix B)

$$\begin{aligned}
d^2\measuredangle f &= \left[\frac{\partial^2 \measuredangle f}{\partial y\,\partial x} - \frac{\partial^2 \measuredangle f}{\partial x\,\partial y} \right] dx \wedge dy \\
&= 0 \ .
\end{aligned} \tag{8.2.6}$$

Note that (8.2.6) is equivalent to the familiar condition that the mixed second partial derivatives of $\measuredangle f$ be equal.

From the Cauchy-Riemann equations it follows that (8.2.6) is also equivalent to the requirement that $\log|f|$ satisfy **Laplace's equation:**

$$\begin{aligned}
\nabla^2 \log|f| &\triangleq \frac{\partial^2 \log|f|}{\partial x^2} + \frac{\partial^2 \log|f|}{\partial y^2} \\
&= 0
\end{aligned} \tag{8.2.7}$$

Definition 8.2.6[Con78]: If D is an open subset of $\mathbb{C}$ then a function $g: D \rightarrow \mathbb{R}$ is *harmonic* if g has continuous second partial derivatives and satisfies Laplace's equation

$$\frac{\partial^2 g}{\partial x^2} + \frac{\partial^2 g}{\partial y^2} = 0 \quad . \tag{8.2.8}$$

■

If D is a simply connected region and $g: D \rightarrow IR$ is harmonic on D, then it may be shown [Con78] that there exists another harmonic function $h: D \rightarrow IR$ such that the function $f: D \rightarrow IR$ defined by $f \triangleq g + jh$ is analytic on D. Such a function (there are infinitely many) is termed a *harmonic conjugate* of g. It is a fact that the real part of any function analytic on a region D is harmonic on D and that the imaginary part of the analytic function is a harmonic conjugate of the real part. In particular, suppose that a transfer function has a logarithm which is analytic on some simply connected region D. It then follows that the phase, $\measuredangle f(s)$, is a harmonic conjugate of the log magnitude, $\log|f(s)|$, on the region D. This observation completes the facts from complex variable theory which we need at present.

8.3. Failure of the Logarithm of a Singular Value to be Harmonic

In Chapter 5 we saw that the concept of gain may be generalized from scalar to multivariable systems using the singular values of a matrix transfer function. By analogy with the scalar case, it is tempting to suppose that each singular value of a matrix transfer function determines the value of a quantity analogous to phase. At the least, the degree-of-freedom argument discussed in Section 8.1 suggests that knowledge of a singular value "gain" should implicitly contain knowledge of some other property of the matrix. These observations motivate us to consider the conjecture that each singular value is the *magnitude* of an analytic function.[25] The results we reviewed in Section 8.2 show that this conjecture can be true only if $\nabla^2 \log\sigma_i = 0$. A general formula for the Laplacian will be derived in Section 10.2. For now, it suffices to show that the conjecture is false by providing a counterexample.

---

[25]As discussed in [DoS81], singular values, being real-valued, cannot themselves be complex analytic.

Consider the matrix

$$M(s) = \begin{bmatrix} s & 0 \\ 2 & s \end{bmatrix} \tag{8.3.1}$$

The singular values of $M(s)$ are

$$\sigma_1 = 1 + \sqrt{|s|^2+1}$$

and (8.3.2)

$$\sigma_2 = -1 + \sqrt{|s|^2+1} \quad ,$$

from which it may readily be verified that

$$\frac{\partial^2 \log\sigma_1}{\partial x^2} + \frac{\partial^2 \log\sigma_1}{\partial y^2} = \left[\frac{1}{\sqrt{|s|^2+1}} \right]^3$$

and (8.3.3)

$$\frac{\partial^2 \log\sigma_2}{\partial x^2} + \frac{\partial^2 \log\sigma_2}{\partial y^2} = -\left[\frac{1}{\sqrt{|s|^2+1}} \right]^3 .$$

Since the right hand sides of (8.3.3) are nonzero, it is clear that log σ_i is not harmonic and that a singular value, in general, cannot be the magnitude of an analytic function.

The example above shows that knowledge of the log of a singular value does not suffice to determine the value of a harmonic conjugate in the same way that scalar gain determines scalar phase. Nevertheless, the degree-of-freedom arguments compel us to pursue the matter a bit further. Consider a matrix transfer function $M(s)$. Using the fact that $M(s)$ is analytic, we shall now show that knowledge of $\sigma_i[M(s)]$ over a region $D \subseteq \mathbb{C}$ does suffice to determine an additional property of the matrix in this region. Additional mathematical development is necessary before we can provide an interpretation of this property and we shall not do so until Chapter 10.

Our first step will be to obtain general expressions for the partial derivatives of $\sigma_i[M(s)]$ with respect to the real and imaginary parts of the complex frequency variable. The following theorem follows easily from a general formula for the derivative of a distinct singular value in[FLC82]. This reference contains rigorous justification of this formula as well as extensions to the case of repeated singular values obtained using the results of Kato[Kat82]. Additional discussion is found in [Fre85].

Theorem 8.3.1[FLC82]: Let $M(s)$ take values in $\mathbb{C}^{n\times n}$ and assume that $M(s)$ is analytic, nonsingular and has distinct singular values at each point of a region $D \subseteq \mathbb{C}$. Then at each point of D

$$\frac{\partial \sigma_i}{\partial x} = \mathrm{Re}\left[v_i^H \frac{\partial M}{\partial x} u_i\right]$$

and (8.3.4)

$$\frac{\partial \sigma_i}{\partial y} = \mathrm{Re}\left[v_i^H \frac{\partial M}{\partial y} u_i\right]$$

where u_i and v_i are, respectively, a pair of right and left singular vectors associated with σ_i.

■

The fact that $M(s)$ is analytic implies that each element of $M(s)$, written in standard coordinates, must satisfy the Cauchy-Riemann equations. This in turn yields

Lemma 8.3.2 Let u and v be vectors in $\mathbb{C}^n$. Then, at each point $s \in \mathbb{C}$ for which $M(s)$ is analytic, it follows that

$$\mathrm{Re}[v^H \frac{\partial M}{\partial x} u] = \mathrm{Im}[v^H \frac{\partial M}{\partial y} u]$$

and (8.3.5)

$$\mathrm{Re}[v^H \frac{\partial M}{\partial y} u] = -\,\mathrm{Im}[v^H \frac{\partial M}{\partial x} u] \quad .$$

■

Proof: The Cauchy-Riemann equations (8.2.1) yield $\frac{\partial M}{\partial x} = \frac{1}{j}\frac{\partial M}{\partial y}$, so that $v^H \frac{\partial M}{\partial x} u =$

$\frac{1}{j} v^H \frac{\partial M}{\partial y} u$. Taking real and imaginary parts yields the result.

■

Observe that equations (8.3.5) are similar to the Cauchy-Riemann equations (8.2.1). Together, equations (8.3.4) and (8.3.5) may be used to show that each singular value of a matrix transfer function does indeed determine another property of the matrix. To see this, consider the one-form $*d\log\sigma_i$ whose value is determined by that of $d\log\sigma_i$ (Appendix B):

$$\begin{aligned} *d\log\sigma_i &= -\frac{\partial\log\sigma_i}{\partial y}dx + \frac{\partial\log\sigma_i}{\partial x}dy \\ &= \frac{-1}{\sigma_i}\mathrm{Re}[v_i^H\frac{\partial M}{\partial y}u_i]dx + \frac{1}{\sigma_i}\mathrm{Re}[v_i^H\frac{\partial M}{\partial x}u_i]dy \\ &= \frac{1}{\sigma_i}\mathrm{Im}[v_i^H\frac{\partial M}{\partial x}u_i]dx + \frac{1}{\sigma_i}\mathrm{Im}[v_i^H\frac{\partial M}{\partial y}u_i]dy \end{aligned} \tag{8.3.6}$$

Note the similarity between (8.3.6) and the differential form (8.2.4) which relates scalar gain and phase. This similarity might tempt one to conjecture that each singular value is associated with a function θ_i such that $*d\log\sigma_i = d\theta_i$. By the Poincaré Lemma (Appendix B) such a function exists if and only if the form (8.3.6) is *exact;* i.e., $d*d\log\sigma_i = 0$. But, since $d*d\log\sigma_i = \nabla^2\log\sigma_i\, dx\wedge dy$, and since our example showed that the Laplacian of $\log\sigma_i$ is not generally zero, it follows that no function θ_i with the desired property can exist. To summarize, it is clear from (8.3.6) that the differential of each singular value contains additional information about the differential of the underlying matrix; i.e., the value of $\mathrm{Im}[v_i^H dMu_i]$. However, the failure of (8.3.6) to be exact implies that this additional information cannot be the differential of another function.

8.4. Summary

The results of this chapter are now summarized. Motivated by an argument based upon degrees of freedom, we have conjectured that each singular value of a matrix transfer function implicitly contains information about another property of the matrix. Since singular values are generalizations of the gain of a scalar transfer function, it seems reasonable to conjecture that a harmonic conjugate can be associated to each singular value in the same way that scalar phase is associated to scalar gain. The counterexample presented in Section 8.3 shows that this conjecture is false. Hence there exists *no* function θ_i such that $\sigma_i e^{j\theta_i}$ is complex analytic.

Despite the failure of $\log\sigma_i$ to be harmonic, we can still show, using the differential form (8.3.6), that each singular value *does* determine another property of the matrix transfer function. We shall devote the next three chapters to determining precisely what this property is. Before embarking on this endeavor, we shall close this chapter by showing that the failure of $\log\sigma_i$ to be harmonic is due to a *directionality* property of a matrix transfer function; specifically, the fact that the singular subspaces change with frequency.

Let $M(s)$, taking values in $\mathbb{C}^{n \times n}$, be analytic and nonsingular and have n distinct singular values over a region $D \subseteq \mathbb{C}$. Then (Appendix A) to each singular value there corresponds a pair of uniquely defined one dimensional left and right singular subspaces, denoted $\mathbf{V}_i$ and $\mathbf{U}_i$, respectively. If these subspaces are constant over D, then to each σ_i a set of constant unit vectors v_i and u_i may be associated[26] so that $v_i \in \mathbf{V}_i$, $u_i \in \mathbf{U}_i$, and $|v_i^H M(s) u_i| = \sigma_i(s)$, $\forall\ s \in D$. Since the function $v_i^H M(s) u_i$ is analytic over D, it follows that a singular value is the magnitude of an analytic function whenever the associated pair of singular subspaces is *constant*. Hence, it is clear that motion of the singular subspaces is responsible for the failure of $\log\sigma_i$ to be harmonic. To explore this phenomenon further, it is necessary to study the mathematical structure of the space in which singular vectors lie.

---

[26] The pair (v_i, u_i) fails to be a pair of singular vectors only in that $v_i^H M(s) u_i$ cannot generally be real valued $\forall\ s \in D$.

CHAPTER 9

STRUCTURE OF THE COMPLEX UNIT SPHERE AND SINGULAR VECTORS

9.1. Introduction

In the previous chapter we argued that complex analyticity should impose structure upon a matrix transfer function analogous to that imposed upon a scalar transfer function by the Bode gain-phase relation. In particular, it seemed reasonable to conjecture that each singular value "gain" is the magnitude of an analytic function and hence determines the value of a function analogous to scalar phase. We showed that this conjecture is generally false. However, if the subspaces spanned by a pair of singular vectors are constant, then the associated singular value is indeed the magnitude of an analytic function.

The preceding observations suggest viewing motion of the singular subspaces with frequency as the *reason* that the associated singular value is not the magnitude of an analytic function. To investigate this matter further, it is convenient to have a mathematical framework within which to describe a pair of singular vectors. The purpose of this chapter is to introduce such a framework. We shall show that a vector in $\mathbb{C}^n$ can be described by specifying its magnitude, the subspace which it spans, and a number analogous to scalar phase.

To provide motivation for the mathematical details, we first show how a complex unit vector may be described by specifying the one-dimensional subspace which it spans and then give a rule for distinguishing among all unit vectors spanning the same subspace. The latter rule is analogous to a definition of phase. We show that it is possible to extend the rule for defining phase in a given subspace to apply to all vectors not orthogonal to the original subspace. It isn't clear, however, whether there exists a single rule applicable to all vectors in

$\mathbb{C}^n$. Neither is it clear how two different rules for defining phase are related on the overlap of their domains of definition.

To answer these and other questions, we introduce some elementary geometric concepts in Section 9.2. We use these ideas to show that the complex unit sphere may be described as a principal fiber bundle; specifically, a circle bundle over complex projective space. This geometric framework allows the concepts of phase and direction to be made precise. In Section 9.3, we shall develop methods of measuring the phase difference between a pair of left and right singular vectors. These measures are generalizations of the notion of phase difference already defined in Chapter 6.

Section 9.4 will be devoted to a method of assigning coordinates to complex projective space. For the case $n = 2$, corresponding to a system with two inputs and two outputs, it is possible to visualize these coordinates as latitude and longitude on a two-dimensional sphere. In Section 9.5, these coordinates will be used to parameterize the singular value decomposition of a matrix $M \in \mathbb{C}^{2\times2}$. This parameterization will prove quite useful in later chapters by allowing us to relate the behavior of the singular values and vectors to the standard components of the matrix.

Section 9.6 summarizes the results of this chapter. In particular, we shall conclude that the action of a linear system upon an input may be decomposed into three parts: a change in magnitude, in direction, and in *phase*. In Chapter 10 the results of this chapter will be used to show how motion of the subspaces spanned by a pair of singular vectors prevents the associated singular value and measure of phase difference from satisfying the Cauchy-Riemann equations.

9.2. Complex Projective Space and Fiber Bundles

In this section we make precise a number of intuitive notions concerning complex unit vectors. Recall that when we speak of the *direction* of a vector $w \in \mathbb{C}^n$, we mean the one-complex-dimensional subspace of $\mathbb{C}^n$ in which it lies. We wish to show that any nonzero vector can be described by specifying its direction, its magnitude, and a number analogous to scalar phase. Since the magnitude of a vector is a familiar concept, we focus our attention on complex *unit* vectors.

It is convenient to introduce a space whose points are the one-dimensional subspaces of $\mathbb{C}^n$. First, we identify $\mathbb{C}^n$ with $\mathbb{R}^{2n}$ in the usual way. The unit sphere in $\mathbb{C}^n$ may then be identified with the sphere $S^{2n-1} \subseteq \mathbb{R}^{2n}$; by mild abuse of notation, the complex unit sphere will also be denoted S^{2n-1}.

Definition 9.2.1 ([Vic73, GrH81]): For each integer n, identify $\mathbb{R}^{2n}$ with $\mathbb{C}^n$ and denote the points of $\mathbb{C}^n$ by $z = (z_1, \ldots, z_n)$. Then the unit sphere $S^{2n-1} \subseteq \mathbb{C}^n$ is defined by

$$S^{2n-1} = \{z \in \mathbb{C}^n : \sum_{i=1}^{n} |z_i|^2 = 1\} \tag{9.2.1}$$

Define an equivalence relation on S^{2n-1} by $z \sim z'$ if and only if there exists a complex number λ with $|\lambda| = 1$ such that $z = \lambda z'$. The quotient space $S^{2n-1}/\sim$ is denoted by $\mathbb{C}P^{n-1}$ and is termed *$(n-1)$-dimensional complex projective space.*

■

The above definition states that the points of $\mathbb{C}P^{n-1}$ are equivalence classes of unit vectors in $\mathbb{C}^n$ and that there is a 1-1 correspondence between the points of $\mathbb{C}P^{n-1}$ and the one-dimensional subspaces of $\mathbb{C}^n$. Given a point $z \in S^{2n-1}$, the corresponding equivalence class will be denoted $[z] \in \mathbb{C}P^{n-1}$. The map taking each unit vector to the appropriate point of $\mathbb{C}P^{n-1}$ is termed the *projection map* and is denoted $\pi : z \rightarrow [z]$. When it is necessary to refer to a point of $\mathbb{C}P^{n-1}$, the notation $\mathbf{Z} \in \mathbb{C}P^{n-1}$ will be used. This notation is consistent with our previous use of boldface letters to denote subspaces of $\mathbb{C}^n$, and the 1-1 correspondence which exists between the points of $\mathbb{C}P^{n-1}$ and the one-dimensional subspaces of $\mathbb{C}^n$.

We shall now show that the vectors in each equivalence class may be distinguished by assigning to each a number $\theta \in [0,2\pi)$. Hence each equivalence class "looks" like the unit circle S^1 with the number θ playing a role analogous to phase.

To illustrate, consider a one-dimensional subspace $\mathbf{W} \subseteq \mathbb{C}^n$ and choose an arbitrary unit vector $w_0 \in \mathbf{W}$. Then, for each nonzero vector $w \in \mathbf{W}$, we may define

$$\theta_{w_0}(w) \triangleq \measuredangle\, w_0^H w \quad . \tag{9.2.2}$$

Restricting the mapping $\theta_{w_0}(\cdot)$ to unit vectors in $\mathbf{W}$ implies that $w_0^H w = \exp[j\,\theta_{w_0}(w)]$ and we obtain the desired 1-1 correspondence between $\mathbf{W} \cap S^{2n-1}$ and S^1.

In the preceding construction, we may consider w_0 as a "phase zero" reference vector and the measure of phase $\theta_{w_0}(w)$ as the "phase difference" between w_0 and w. Clearly, the definition (9.2.2) may be extended to measure the phase difference between w_0 and any vector in $\mathbb{C}^n$ lying in a direction not orthogonal to w_0. Equivalently, (9.2.2) gives a method of defining a reference vector in each non-orthogonal direction; i.e., given $\mathbf{Z}$ not orthogonal to $\mathbf{W}$, define the reference vector $z_0 \in \mathbf{Z}$ to be the unique unit vector with $w_0^H z_0$ real and positive. Each unit vector $z \in \mathbf{Z}$ may then be written as $z = z_o \exp[j\,\theta_{w_0}(w)]$.

Given the interpretation of phase difference between vector-valued signals presented in Chapter 6, it is reasonable that the use of (9.2.2) to measure phase cannot be extended to all of $\mathbb{C}^n$. However, given a basis $\{w_i \ ; \ i = 1, \ldots, n\}$, one can use each basis vector to construct a measure of phase, θ_{w_i}, as in (9.2.2). It follows easily that each vector in $\mathbb{C}^n$ is guaranteed to lie within the domain of definition of at least one of the θ_{w_i}. This procedure motivates asking how these different measures of phase are related on the overlap of their domains of definition. It also leaves as an open question the possible existence of a measure of phase applicable to all nonzero vectors in $\mathbb{C}^n$.

To make precise the concepts of vector phase and direction, it is necessary to have a mathematical description of the complex unit sphere. Once such a description is obtained, the questions raised in the preceding paragraph may be answered. First, a number of definitions are needed. These are all from [Ble81], to which the reader is referred for a complete discussion. Another useful reference is[Spi79].

Definition 9.2.2: A *principal fiber bundle* consists of a manifold P (called the *total space*), a Lie group G, a *base manifold* M, and a *projection map* $\pi : P \to M$ such that conditions (A), (B) and (C) are satisfied.

(A) The group G acts freely and differentiably on P to the right.

(B) The map $\pi : P \to M$ is onto and $\pi^{-1}(\pi(p)) = \{pg : g \in G\}$. If $x \in M$, then $\pi^{-1}(x)$ is called the *fiber* above x.

(C) For each $x \in M$, there is an open set N_α with $x \in N_\alpha$ and a diffeomorphism $T_\alpha : \pi^{-1}(N_\alpha) \to N_\alpha \times G$ of the form $T_\alpha(p) = (\pi(p), s_\alpha(p))$, where $s_\alpha : \pi^{-1}(N_\alpha) \to G$ has the property that $s_\alpha(pg) = s_\alpha(p)g$ for all $g \in G$, $p \in \pi^{-1}(N_\alpha)$. The map T_α is called a *local trivialization*.

■

Definition 9.2.3: Let $T_\alpha : \pi^{-1}(N_\alpha) \to N_\alpha \times G$ and $T_\beta : \pi^{-1}(N_\beta) \to N_\beta \times G$ be two local trivializations of a principal fiber bundle $\pi : P \to M$ with group G. The *transition function* from T_α to T_β is the map $g_{\alpha\beta} : N_\alpha \cap N_\beta \to G$ defined, for $x = \pi(p) \in N_\alpha \cap N_\beta$, by $g_{\alpha\beta}(x) = s_\alpha(p)[s_\beta(p)]^{-1}$. Note that $g_{\alpha\beta}(x)$ is independent of the choice of $p \in \pi^{-1}(x)$ because

$$s_\alpha(pg)[s_\beta(pg)]^{-1} = s_\alpha(p)g\,[s_\beta(p)g]^{-1} = s_\alpha(p)[s_\beta(p)]^{-1} \quad .$$

The following properties hold:

(*i*) $g_{\alpha\alpha}(y) = e$ for all $y \in N_\alpha$, where e is the identity element of the group G.

(*ii*) $g_{\beta\alpha}(y) = [g_{\alpha\beta}(y)]^{-1}$ for all $y \in N_\alpha \cap N_\beta$

(*iii*) $g_{\alpha\beta}(y)g_{\beta\gamma}(y)g_{\gamma\alpha}(y) = e$ for all $y \in N_\alpha \cap N_\beta \cap N_\gamma$.

■

The transition functions describe how the products $N_\alpha \times G$, $N_\beta \times G$, etc. glue together to form the total space P. Thus P may be considered as the space obtained from the disjoint union $(N_\alpha \times G) \cup (N_\beta \times G) \cup \cdots$ by identifying the point $(x,g) \in N_\alpha \times G$ with $(x,\hat{g}) \in N_\beta \times G$ if $g = g_{\alpha\beta}(x)\hat{g}$.

Definition 9.2.4: A *local section* of a principal fiber bundle $\pi : P \to M$ with group G is a map $\rho_\alpha : N_\alpha \to P$, where N_α is an open subset of M, such that $\pi \circ \rho_\alpha = 1_\alpha \triangleq$ the identity function on N_α.

■

Theorem 9.2.5 [Ble81]: There is a natural correspondence between local sections and local trivializations.

■

Proof: If $\rho_\alpha : N_\alpha \to P$ is a local section, then define $T_\alpha = \pi^{-1}(N_\alpha) \to N_\alpha \times G$ by $T_\alpha(\rho_\alpha(x)g) = (x,g)$. Conversely, given a local trivialization $T_\alpha : \pi^{-1}(N_\alpha) \to N_\alpha \times G$, define a local section $\rho_\alpha : N_\alpha \to P$ by $\rho_\alpha(x) = T_\alpha^{-1}(x,e)$, where e is the identity element of G.

■

Definition 9.2.6: If T_α is a local trivialization with $N_\alpha = M$ (i.e., $T_M : P \to M \times G$), then T_M is called a *global trivialization*, and the principal fiber bundle is called *trivial* if such a T_M exists.

■

In our case the total space is the unit sphere in $\mathbb{C}^n$, identified with $S^{2n-1} \subsetneq \mathbb{R}^{2n}$, the base manifold is complex projective space $\mathbb{C}P^{n-1}$, and the Lie group is the scalar-valued unitary group

$$Un(1) \triangleq \{e^{ja} : a \in \mathbb{R}\} \quad . \tag{9.2.3}$$

Note that $Un(1)$ may be identified with the unit circle S^1. The projection map is the map taking a unit vector in $\mathbb{C}^n$ to its equivalence class under the equivalence relation given following Definition 9.2.1.

It is straightforward to verify that conditions (A)-(C) are satisfied. Condition (A) means, for the present purpose, that for each element $g = e^{ja} \in Un(1)$, there exists a nicely behaved mapping $R_g: S^{2n-1} \to S^{2n-1}$ defined by $R_g(z) = ze^{ja}$, $z \in S^{2n-1}$. Condition (B) means that there exists a 1-1 correspondence between each fiber of the bundle and the Lie group $Un(1)$. The identification of $Un(1)$ with S^1 motivates referring to the complex unit sphere as a *circle bundle over complex projective space*. Indeed, this is just an alternate way of stating that the collection of unit vectors in a one dimensional subspace of $\mathbb{C}^n$ may be distinguished by assigning to each a value of phase $\theta \in [-\pi,\pi)$. Finally, we can show that condition (C) is satisfied by using a set of basis vectors for $\mathbb{C}^n$ to construct the maps T_α. Let $\{w_i \; ; \; i=1, \ldots, n\}$ denote such a basis and define the subset $N_{w_i} \subsetneq \mathbb{C}P^{n-1}$ as the collection of all directions in $\mathbb{C}^n$ not orthogonal to that spanned by w_i. Equivalently, N_{w_i} has the characterization

$$\pi^{-1}(N_{w_i}) = \{w \in S^{2n-1} : w_i^H w \neq 0\} \quad . \tag{9.2.4}$$

Next, the measure of phase (9.2.2) may be used to construct a set of local trivializations:

$$\begin{aligned} T_{w_i} &: \pi^{-1}(N_{w_i}) \to N_{w_i} \times S^1 \\ T_{w_i}(w) &= ([w], \exp[j\theta_{w_i}(w)]) \end{aligned} \tag{9.2.5}$$

where $\theta_{w_i}(w)$ is given by (9.2.2). It is straightforward to verify that T_{w_i} is a diffeomorphism; i.e., a differentiable mapping from $\pi^{-1}(N_{w_i})$ to $N_{w_i} \times S^1$ which is 1-1 and onto and whose inverse is also differentiable. Note that $\theta_{w_i}(we^{ja}) = a + \theta_{w_i}(w)$; hence, the property $s_\alpha(pg) = s_\alpha(p)g$ is satisfied.

Condition (C) may be interpreted as stating that *locally* it is possible to define the phase of a vector signal by measuring the phase difference between it and a reference signal. Hence each of the local trivializations may be viewed as a local measure of phase. It is natural to inquire whether a globally applicable measure of phase exists. The key to answering this question lies in the application of some mathematical techniques available to us because of the correspondence between local trivializations and local definitions of phase. The answer, as might be suspected given the interpretation of phase in Chapter 6, is negative. We shall show this by demonstrating the non-existence of a global trivialization.

Indeed, suppose that a global trivialization existed. Then, by definition, there would exist a homeomorphism $T : S^{2n-1} \to \mathbb{C}P^{n-1} \times S^1$. (A homeomorphism is a continuous 1-1 onto mapping with a continuous inverse [Mun75].) A well-known result from topology is that homeomorphisms preserve the property of being simply connected [Mun75]. The existence of the map T would therefore imply that either (a) both S^{2n-1} and $\mathbb{C}P^{n-1} \times S^1$ are simply connected or (b) neither space has this property. It may be shown [Hu59] that the unit sphere S^m is simply connected if and only if $m > 1$. Hence the unit sphere $S^{2n-1} \subseteq \mathbb{C}^n$, $n > 1$, is simply connected while S^1 (and thus the product space $\mathbb{C}P^{n-1} \times S^1$) is not. This contradiction shows that no map T with the desired properties exists. The interpretation of a trivialization of S^{2n-1} as a means of defining phase thus implies that no single definition of phase applicable to all nonzero vectors in $\mathbb{C}^n$ can exist.

Just as a local trivialization makes precise the idea of defining "phase" of a vector, a local section makes precise the idea of choosing a phase-zero reference vector in each direction. To illustrate, consider the local trivialization T_{w_i}. Let $\mathbf{Z}$ denote a one-dimensional subspace of $\mathbb{C}^n$ (and the corresponding point of $\mathbb{C}P^{n-1}$). If $\mathbf{Z}$ is not orthogonal to w_i, then one can choose a reference vector $\rho_{w_i}(\mathbf{Z})$ by choosing an arbitrary unit vector $z \in \mathbf{Z}$ and using the rule

$$\begin{aligned} \rho_{w_i}(\mathbf{Z}) &= T_{w_i}^{-1}(\mathbf{Z}\ ,\ 1) \\ &= \exp[-j\,\theta_{w_i}(z)] \cdot z \end{aligned} \qquad (9.2.6)$$

Equivalently, $\rho_{w_i}(\mathbf{Z})$ is the unique unit vector in $\mathbf{Z}$ such that $w_i^H \rho_{w_i}(\mathbf{Z})$ is real and positive.

Given a basis $\{w_i\ ;\ i=1, \ldots, n\}$ for $\mathbb{C}^n$, each unit vector $z \in \mathbb{C}^n$ lies in the domain of definition of at least one trivialization T_{w_i} (9.2.5). Hence, using an appropriate local section and trivialization, any unit vector may be written

$$z = \rho_{w_i}([z])\exp[j\,\theta_i(z)] \quad . \qquad (9.2.7)$$

Note in particular that the vector $\rho_{w_i}(\cdot)$ is a function *only* of the subspace spanned by z.

The concepts of local trivialization and local section of a principal fiber bundle are very useful in formalizing the ideas of measuring phase of a vector and defining phase zero reference vectors which we discussed at the beginning of this section. In particular, the transition functions show how the different measures of phase obtained using a set of basis vectors for $\mathbb{C}^n$ are related on the overlap of their domains of definition. Specifically, given the local trivializations T_{w_i} and T_{w_k} and a vector $p \in \pi^{-1}(N_{w_i}) \cap \pi^{-1}(N_{w_k})$, we may write

$$\exp(j\,\theta_{w_i}(p)) = g_{w_i w_k}([p])\exp(j\,\theta_{w_k}(p)) \quad , \tag{9.2.8}$$

where the transition function $g_{w_i w_k}([p])$ may be evaluated by choosing an *arbitrary* vector $\hat{p} \in [p]$ and computing

$$g_{w_i w_k}([p]) = \exp[j\,\sphericalangle\overline{(w_k^H\hat{p})}(w_i^H\hat{p})] \quad . \tag{9.2.9}$$

The local sections corresponding to these local trivializations are also related by the transition functions:

$$\rho_{w_k}([p]) = \rho_{w_i}([p])\, g_{w_i w_k}([p]) \quad . \tag{9.2.10}$$

These ideas will now be illustrated using a set of local trivializations defined by the standard basis vectors for $\mathbb{C}^n$, denoted $\{e_i = [0\ 0 \cdots \overset{i}{1} \cdots 0]^T;\ i = 1, \ldots, n\}$. Consider a unit vector $z = \sum_{i=1}^{n} z_i e_i$ and assume that $z_i \neq 0$. Then z lies in the domain of definition of the local trivialization T_{e_i} defined by (9.2.5) and

$$T_{e_i}(z) = ([z]\ ,\ \exp[j\,\theta_{e_i}(z)])$$

where (9.2.11)

$$\begin{aligned}\theta_{e_i}(z) &= \sphericalangle\, e_i^H z \\ &= \sphericalangle\, z_i \quad .\end{aligned}$$

We see that T_{e_i} measures the phase of a vector by measuring the phase of its ith component, provided that this component is nonzero. If the direction $\mathbf{Z}$ is not orthogonal to e_i, then it is possible to define a reference vector in $\mathbf{Z}$ using a local section and an *arbitrarily* chosen vector $\hat{z} \in \mathbf{Z}$:

$$\rho_{e_i}(\mathbf{Z}) = \exp(-j \measuredangle \hat{z}_i) \cdot \hat{z} \quad . \tag{9.2.12}$$

As an example, consider

$$z = \begin{bmatrix} \frac{\sqrt{2}}{2} \\ j\frac{\sqrt{2}}{2} \end{bmatrix}$$

Then, writing

$$z = \rho_{e_i}([z])\exp(j\ \theta_{e_i}(z))$$

yields

$$(i=1) \qquad z = \begin{bmatrix} \frac{\sqrt{2}}{2} \\ j\frac{\sqrt{2}}{2} \end{bmatrix} \cdot \exp(0)$$

$$(i=2) \qquad z = \begin{bmatrix} -j\frac{\sqrt{2}}{2} \\ \frac{\sqrt{2}}{2} \end{bmatrix} \cdot \exp(j\frac{\pi}{2})$$

The transition functions $g_{e_i e_k}(\mathbf{Z})$, may be computed from (9.2.9):

$$g_{e_i e_k}(\mathbf{Z}) = \exp[j \measuredangle \bar{\hat{z}}_k\ \hat{z}_i] \tag{9.2.13}$$

where $\hat{z}_k$ and $\hat{z}_i$ are the kth and i^{th} components of an arbitrarily chosen vector $z \in \mathbf{Z}$. For our example,

$$g_{e_1 e_2}(\mathbf{Z}) = \exp(-j\pi/2)$$

and

$$g_{e_2 e_1}(\mathbf{Z}) = \exp(j\pi/2) \quad .$$

It may be verified that (9.2.8) and (9.2.10) are satisfied.

9.3. Phase Difference between a Pair of Singular Vectors

The purpose of this section is to use local trivializations to obtain measures of phase difference between a pair of left and right singular vectors.

Let $M \in \mathbb{C}^{n \times n}$ have a singular value σ with multiplicity one.[27] Then associated to σ is a pair of uniquely defined one-dimensional left and right singular subspaces, denoted V and U, respectively. As discussed in Appendix A, the right singular vector may be chosen as any unit vector $u \in$ U and the left singular vector is then uniquely defined by the equation $v = Mu/\sigma$. Hence, if (u,v) is a valid singular vector pair, then so is $(\hat{u},\hat{v}) = e^{ja}(u,v)$, $\forall \ a \in IR$. Evidently, then, the "phase" of a singular vector is not a well-defined quantity. Nevertheless, appropriate measures of *phase difference* between the singular vectors *are* well defined. We have already seen one measure of phase difference in Chapter 6, and the following discussion will yield a general method for obtaining others.

One way of measuring phase difference between a pair of vectors is to compute the difference in phase between each and a third vector. Equivalently, suppose that both vectors lie within the domain of definition of the same local trivialization. Then we can define the phase difference by

$$\delta\theta_{w_i}(u,v) = \theta_{w_i}(v) - \theta_{w_i}(u) \tag{9.3.1}$$

where $\theta_{w_i}(\cdot)$ is defined by (9.2.2), and we renormalize, if necessary, so that $\delta\theta_{w_i}(u,v) \in [-\pi,\pi)$. Clearly, $\delta\theta_{w_i}(e^{ja}u, e^{ja}v) = \delta\theta_{w_i}(u,v)$ for all $a \in IR$, so that this measure of phase difference is indeed independent of the choice of singular vector pair.

---

[27] For notational simplicity, the subscript ordering the singular values will be suppressed when possible.

The transition functions (9.2.9) may be used to derive functions showing how the values of phase difference obtained using different local trivializations are related.

Lemma 9.3.1: Suppose that the vectors u and v each lie within the domain of definition of the local trivializations T_{w_i} and T_{w_k}. Then the associated measures of phase difference, $\delta\theta_{w_i}(u,v)$ and $\delta\theta_{w_k}(u,v)$, satisfy the identity

$$\exp[j\,\delta\theta_{w_i}(u,v)] = \delta g_{w_i w_k}([u],[v])\exp[j\,\delta\theta_{w_k}(u,v)] \quad , \tag{9.3.2}$$

where $\delta g_{w_i w_k}([u],[v])$ is the δ*–transition function* defined by

$$\delta g_{w_i w_k}([u],[v]) \overset{\Delta}{=} g_{w_i w_k}([v])g_{w_k w_i}([u]) \quad . \tag{9.3.3}$$

■

Proof: By definition, the transition functions satisfy

$$\exp(j\,\theta_{w_i}(v)) = g_{w_i w_k}([v])\exp(j\,\theta_{w_k}(v))$$

and (9.3.4)

$$\exp(j\,\theta_{w_i}(u)) = g_{w_i w_k}([u])\exp(j\,\theta_{w_k}(u))$$

which implies

$$\exp(j\,\delta\theta_{w_i}(u,v)) = g_{w_i w_k}([v])\overline{g_{w_i w_k}([u])}\,\exp(j\,\delta\theta_{w_k}(u,v)) \quad . \tag{9.3.5}$$

The fact that $\overline{g_{w_i w_k}([u])} = g_{w_k w_i}([u])$ yields the desired result.

■

Note that, once the local trivializations are specified, the value of the δ-transition function may be computed using only knowledge of V and U, the left and right singular subspaces. Indeed, this function may be computed by choosing *arbitrary* vectors $\hat{u} \in$ U and $\hat{v} \in$ V and calculating

$$\begin{aligned} \delta g_{w_i w_k}(\mathrm{U},\mathrm{V}) &= g_{w_i w_k}(\mathrm{V}) g_{w_k w_i}(\mathrm{U}) \\ &= \exp[j \sphericalangle(\overline{w_k^H \hat{v}})(w_i^H \hat{v})] \exp[-j \sphericalangle(\overline{w_k^H \hat{u}})(w_i^H \hat{u})] \quad . \end{aligned} \tag{9.3.6}$$

Local trivializations defined using the standard basis vectors will prove useful in parameterizing complex matrices and in relating the singular vectors of a matrix to the components of the matrix written in standard coordinates. When a local trivialization is defined using e_i, the ith standard basis vector, the resulting measure of phase difference is equal to the difference between the phases of the ith components of the singular vectors:

$$\delta\theta_{e_i}(u,v) = \sphericalangle e_i^H v - \sphericalangle e_i^H u \quad . \tag{9.3.7}$$

Since we shall use such measures of phase difference often in later sections, it will prove convenient to adopt the following notational shorthand for (9.3.7):

$$\delta\theta_i(u,v) \triangleq \delta\theta_{e_i}(u,v) \quad . \tag{9.3.8a}$$

When it is necessary to distinguish among the pairs of singular vectors, (9.3.8a) will be further abbreviated

$$\delta\theta_i(k) = \delta\theta_i(u_k,v_k) \quad . \tag{9.3.8b}$$

Hence $\delta\theta_i(k)$ is the difference in phase between the ith components (in the standard basis) of the kth pair of singular vectors.

As we have discussed previously, the individual singular vectors are not well-defined quantities. However, it will sometimes prove convenient to have a convention for choosing one member of a pair of singular vectors, after which the other vector is well defined. The preceding development suggests that one method for doing this is to choose, e.g., the right singular vector to be the phase-zero reference vector in the local trivialization T_{w_i}. This choice yields the pair of singular vectors

$$\begin{aligned} u &\triangleq \rho_{w_i}(\mathrm{U}) \\ v &\triangleq \rho_{w_i}(\mathrm{V})\exp(j\,\delta\theta_{w_i}(u,v)) \quad . \end{aligned} \tag{9.3.9}$$

For example, the local trivialization defined using the ith standard basis vector yields a right singular vector whose i^{th} component, in standard coordinates, is real and positive.

When the singular value decomposition in question is that obtained from a stable matrix transfer function evaluated at $j\omega_0$, the phase difference $\delta\theta_i(u,v)$ determines how the ith components of an input $u(t) = ue^{j\omega_0 t}$ and the steady state output $v(t) = ve^{j\omega_0 t}$ interfere when added together. Although useful for certain purposes, this measure of phase difference does not have as appealing a physical interpretation as that discussed in Chapter 6, which determines how the *total* input and resulting output interfere. The latter measure of phase difference can be obtained in the present context by measuring the phase of each member of a pair of singular vectors using a trivialization defined from an (arbitrarily chosen) right singular vector:

$$\begin{aligned} \delta\theta_u(u,v) &= \theta_u(v) - \theta_u(u) \\ &= \measuredangle u^H v - 0 \end{aligned} \tag{9.3.10}$$

The notation $\delta\theta_u(u,v)$ is a bit misleading, since the subscript u might lead one to believe that this value of phase difference depends upon the choice of the singular vector u, when in fact it does not. For this reason, and also to maintain notational consistency with Chapter 6, we introduce an alternate notation for (9.3.10):

$$\Delta\theta(u,v) \triangleq \delta\theta_u(u,v) \quad . \tag{9.3.11a}$$

By analogy with (9.3.8), when it becomes necessary to distinguish among the pairs of singular vectors, (9.3.11a) will be further abbreviated

$$\Delta\theta(k) = \Delta\theta(u_k, v_k) \tag{9.3.11b}$$

In later chapters we shall derive differential equations which must be satisfied by the phase difference between a pair of singular vectors of a matrix transfer function. In doing so, it will be most convenient to define phase difference using a *constant* local trivialization, rather than one which is frequency dependent, as in (9.3.10-11). There is no loss of generality in doing this, because we can always use δ-transition functions to transform between (9.3.10-11) and other measures of phase difference. These δ-transition functions, denoted $\delta g_{uw_i}(u,v)$, may be computed directly from Lemma 9.3.1. To emphasize that these functions do not depend upon the choice of right singular vector, we prefer the notation $\delta g_{\Delta w_i}(u,v)$, and note that they may be computed using the following shortcut.

Corollary 9.3.2.: Consider U and V, a pair of right and left singular subspaces, and $u \in$ U, $v \in$ V, a pair of right and left singular vectors. Suppose that $u^H v \neq 0$, so that $\Delta\theta(u,v)$ in (9.3.11) is defined, and that u and v each lie in the domain of definition of the local trivialization T_{w_i} (9.2.5). Then the corresponding measures of phase difference (9.3.10-11) and (9.3.1) are related by

$$\exp[j\Delta\theta(u,v)] = \delta g_{\Delta w_i}(\mathrm{U},\mathrm{V})\exp[j\delta\theta_{w_i}(u,v)] \tag{9.3.12}$$

where

$$\delta g_{\Delta w_i}(\mathrm{U},\mathrm{V}) \triangleq \exp\{j\measuredangle[\rho_{w_i}(\mathrm{U})]^H[\rho_{w_i}(\mathrm{V})]\} \quad . \tag{9.3.13}$$

■

Proof: With no loss of generality, u and v can be chosen as in (9.3.9). It follows that

$$\begin{aligned} u^H v &= |u^H v|\exp[j\Delta\theta(u,v)] \\ &= [\rho_{w_i}(\mathrm{U})]^H \rho_{w_i}(\mathrm{V})\exp[j\delta\theta_{w_i}(u,v)] \quad . \end{aligned}$$

Since $|u^H v| > 0$ by assumption, the result follows.

■

Equations (9.3.12)-(9.3.13) show that the difference between $\Delta\theta(u,v)$ and $\delta\theta_{w_i}(u,v)$ is equal to the phase difference between those vectors in the subspaces U and V which have phase zero when measured in the trivialization T_{w_i}.

9.4. Coordinates on Complex Projective Space

In Section 9.2 we saw that the complex unit sphere could be described by assigning to each unit vector the corresponding point of complex projective space plus an additional coordinate analogous to scalar phase. Note that $2n - 2$ real coordinates are needed to distinguish among the points of $\mathbb{C}P^{n-1}$. The purpose of this section is to describe a method of assigning these coordinates. We shall discuss in detail the case $n = 2$, corresponding to a system with two inputs and outputs, and then indicate how the procedure may be extended to the general case.

First, consider the unit sphere in $\mathbb{C}^2$. We shall describe a method of identifying the equivalence classes from Definition 9.2.1 with the points of the unit sphere $S^2 \subseteq \mathbb{R}^3$. Our procedure is similar to one described in [AuM77] and has the advantage of allowing us to visualize points of $\mathbb{C}P^1$ by thinking of latitude and longitude on S^2.

Let the points of the unit sphere in $\mathbb{C}^2$ be written in standard coordinates,[28] $z = [z_1 \; z_2]^T$, and consider the mapping

$$s(z) = \frac{z_2}{z_1} \quad . \tag{9.4.1}$$

---

[28]It is not necessary to assume that z is expressed in the standard basis; any other basis will do as well.

Note that $s(z)$ is a function only of $[z]$, the subspace spanned by z. Hence (9.4.1) establishes a 1-1 correspondence between the points of $\mathbb{C}P^1$ and those of the extended complex plane.

By mapping the extended complex plane onto the unit sphere $S^2 \subseteq \mathbb{R}^3$, it is thus possible to place the points of $\mathbb{C}P^1$ into 1-1 correspondence with the points of $S^2 \subseteq \mathbb{R}^3$. Indeed, consider the coordinates ϕ and ψ defined by

$$\phi \triangleq \arctan \left| \frac{z_2}{z_1} \right| \quad , \quad 0 \leq \phi \leq \pi/2$$

and, if $0 < \phi < \pi/2$, (9.4.2)

$$\psi \triangleq \sphericalangle \frac{z_2}{z_1} \quad , \quad -\pi < \psi \leq \pi \ .$$

As shown in Figure 9.4.1, this set of coordinates on $\mathbb{C}P^1$ is analogous to latitude and longitude. Note that the subspace spanned by $e_1 = [1\ 0]^T$ maps to the North pole (N) and that the subspace spanned by $e_2 = [0\ 1]^T$ maps to the South pole (S). Hence the local trivialization T_{e_1} is defined on unit vectors above points of $\{S^2 - S\}$, while T_{e_2} is defined on unit vectors above points of $\{S^2 - N\}$.

Each point of $\mathbb{C}P^1$ may be described by specifying a set of coordinates (ϕ,ψ), and it is instructive to obtain expressions for the local sections and transition functions using these coordinates. The local sections (9.2.6) defined using the standard basis vectors may be written

$$\rho_{e_1}(\phi,\psi) = \begin{bmatrix} \cos\phi \\ \sin\phi \ e^{j\psi} \end{bmatrix} \tag{9.4.3a}$$

$$\rho_{e_2}(\phi,\psi) = \begin{bmatrix} \cos\phi \ e^{-j\psi} \\ \sin\phi \end{bmatrix} \tag{9.4.3b}$$

Using (9.4.3), any unit vector $z = [z_1\ z_2]^T$ with $z_1 z_2 \neq 0$ may be written using either local

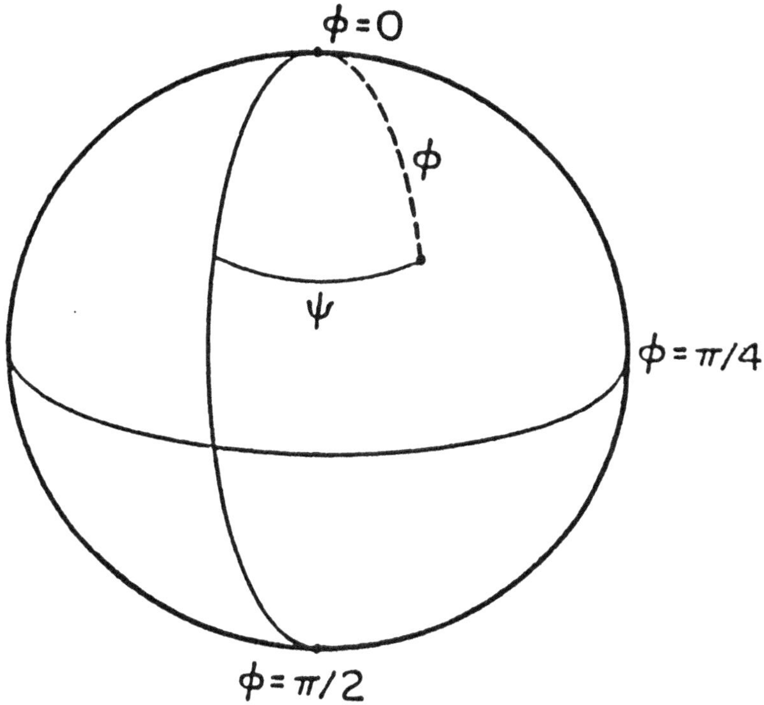

Figure 9.4.1: Coordinates on $\mathbb{C}P^1$

trivialization T_{e_1} or T_{e_2} as

$$z = \begin{bmatrix} \cos\phi \\ \sin\phi \;\; e^{j\psi} \end{bmatrix} e^{j\theta_{e_1}(z)} \quad \text{using } T_{e_1}$$

or (9.4.4)

$$z = \begin{bmatrix} \cos\phi \;\; e^{-j\psi} \\ \sin\phi \end{bmatrix} e^{j\theta_{e_2}(z)} \quad \text{using } T_{e_2} \quad .$$

The transition functions are

$$g_{e_1e_2}(\phi,\psi) = \exp(-j\psi)$$

and (9.4.5)

$$g_{e_2e_1}(\phi,\psi) = \exp(j\psi) \quad .$$

Note that $\theta_{e_2}(z) = \theta_{e_1}(z) + \psi$, as required by Definition 9.2.3.

Expressing the unit sphere in $\mathbb{C}^2$ as a circle bundle over $\mathbb{C}P^1$ is known as the Hopf fibration [Hu59]. This construction is very appealing in that it allows one to visualize the two-dimensional complex unit sphere. In higher dimensions, visualization is no longer possible; however, a mathematical procedure for constructing the space $\mathbb{C}P^n$ from $\mathbb{C}P^{n-1}$ is available ([GrH81], pp.114-115). For our purposes, it will suffice to indicate how coordinates for $\mathbb{C}P^n$ may be obtained from those for $\mathbb{C}P^{n-1}$. Each increase in the value of n requires an additional pair of coordinates analogous to latitude and longitude. We illustrate by examining the resulting local section obtained from the trivialization T_{e_1}. For[29]

$$n = 2 \qquad \rho_{e_1}(\phi_1,\psi_1) = \begin{bmatrix} \cos\phi_1 \\ \sin\phi_1 \ e^{j\psi_1} \end{bmatrix}$$

$$n = 3 \qquad \rho_{e_1}(\phi_1,\psi_1,\phi_2,\psi_2) = \begin{bmatrix} \cos\phi_1 \\ \sin\phi_1 \ \cos\phi_2 \ e^{j\psi_1} \\ \sin\phi_1 \ \sin\phi_2 \ e^{j\psi_2} \end{bmatrix}$$

$$n = 4 \qquad \rho_{e_1}(\phi_1,\psi_1,\phi_2,\psi_2,\phi_3,\psi_3) = \begin{bmatrix} \cos\phi_1 \\ \sin\phi_1 \ \cos\phi_2 \ e^{j\psi_1} \\ \sin\phi_1 \ \sin\phi_2 \ \cos\phi_3 \ e^{j\psi_2} \\ \sin\phi_1 \ \sin\phi_2 \ \sin\phi_3 \ e^{j\psi_3} \end{bmatrix}$$

where $\phi_i \in [0,\pi/2]$ and, if defined, $\psi_i \in (-\pi,\pi]$. Note that the relative magnitudes of the components of the vector are determined by the "latitude" angles ϕ_i, while the phase difference between the $(i+1)^{st}$ and the 1st components is determined by the "longitude" angle ψ_i. It is straightforward to derive formulas analogous to (9.4.5) for the transition functions in these higher dimensional cases; unlike the case $n = 2$, these transition functions will generally depend upon both the latitude *and* longitude coordinates of the subspace in question.

Since the transition functions may be expressed in terms of the latitude and longitude coordinates, it is clear that the δ-transition functions may be expressed in these coordinates also. As an example, we shall compute the δ-transition function (9.3.13) in the special case that $w_i = e_i \in \mathbb{C}^2$. This class of transition functions will prove important in later chapters, where we shall be primarily concerned with the measure of phase difference $\Delta\theta(u,v)$ (9.3.11) and need to relate this to the phase difference between two of the components of v and u

[29]In the case n=2, we shall usually suppress the subscript "1".

written in standard coordinates.

By applying (9.4.3a) to each of the singular subspaces, it follows that the phase-zero reference vectors corresponding to the local trivialization T_{e_1} may be written

$$\begin{aligned} \rho_{e_1}(\phi(V),\psi(V)) &= \begin{bmatrix} \cos\phi(V) \\ \sin\phi(V)\ \ \exp[j\,\psi(V)] \end{bmatrix} \\ \rho_{e_1}(\phi(U),\psi(U)) &= \begin{bmatrix} \cos\phi(U) \\ \sin\phi(U)\ \ \exp[j\,\psi(U)] \end{bmatrix} . \end{aligned} \tag{9.4.6}$$

It follows readily from (9.3.13) that

$$\delta g_{\Delta e_1}(U,V) = \exp\left\{ j\angle \left[\cos\phi(U)\cos\phi(V) + \sin\phi(U)\sin\phi(V)\exp[j(\psi(V) - \psi(U))] \right] \right\} . \tag{9.4.7a}$$

A similar analysis using (9.4.3b) yields

$$\delta g_{\Delta e_2}(U,V) = \exp\left\{ j\angle \left[\cos\phi(U)\cos\phi(V)\exp[-j(\psi(V) - \psi(U))] + \sin\phi(U)\sin\phi(V) \right] \right\} . \tag{9.4.7b}$$

The distinction between these δ-transition functions is, of course, that (9.4.7a) relates $\Delta\theta(u,v)$ to the phase difference between the 1st components of u and v, written in the standard basis, while (9.4.7b) does the same for the 2nd components of these vectors.

Intuitively, the measures of phase difference $\Delta\theta(u,v)$ and $\delta\theta_{e_i}(u,v)$ should be approximately equal if at least one of the singular subspaces is aligned sufficiently closely with one of the standard basis directions $[e_i]$. By definition (9.4.2), the closer the latitude coordinate $\phi(V)$ is to zero, the more closely aligned the left singular subspace V will be with $[e_1]$. Hence, if $\phi(V) = 0$, then $\sin\phi(V) = 0$ and $\delta g_{\Delta e_1}(U,V) = 1$. Hence, $\Delta\theta(u,v) = \delta\theta_{e_1}(U,V)$ and our intuition is verified. A similar analysis applies to the right singular subspace. On the other hand, if $\phi(V) = \pi/2$ or $\phi(U) = \pi/2$, then the corresponding subspace is approximately aligned with $[e_2]$. It follows that $\delta g_{\Delta e_2}(U,V) \approx 1$ and $\delta\theta(u,v) \approx \delta\theta_{e_2}$. Again, our intuition is verified. These observations will prove useful in Chapter 12, where we shall analyze an example whose right singular subspaces are aligned with the standard basis directions.

9.5. Parameterization of Systems with Two Inputs and Outputs

The purpose of this section is to derive a parameterization of the transfer function of a system with two inputs and outputs. In so doing, we shall use the latitude and longitude angles discussed in the previous section. We shall use this parameterization in discussing the example in Chapter 12. A similar parameterization has been discussed by Hung and MacFarlane [HuM82].

Let $L \in \mathbb{C}^{2\times 2}$ be a transfer function evaluated at a frequency of interest. The singular value decomposition of L may be written as

$$L = V \Sigma U^H = [v_1 | v_2] \begin{bmatrix} \sigma_1 & 0 \\ 0 & \sigma_2 \end{bmatrix} \begin{bmatrix} u_1^H \\ \hline u_2^H \end{bmatrix} , \qquad (9.5.1)$$

where we assume that $\sigma_1 > \sigma_2 > 0$. Suppose that the ith left singular vector lies within the domain of definition of the local trivialization T_{e_i}. Then this trivialization may be used to define the phase of v_i and the results from Section 9.4 can be used to obtain latitude and longitude angles describing each subspace V_i. Following (9.4.4), this procedure yields

$$v_1 = \rho_{e_1}(V_1)\exp[j\theta_{e_1}(v_1)] = \begin{bmatrix} \cos\phi(V_1) \\ \sin\phi(V_1)\exp[j\psi(V_1)] \end{bmatrix} \exp[j\theta_{e_1}(v_1)]$$

and (9.5.2)

$$\nu_2 = \rho_{e_2}(\mathbf{V}_2)\exp[j\theta_{e_2}(\nu_2)] = \begin{bmatrix} \cos\phi(\mathbf{V}_2) & \exp[-j\psi(\mathbf{V}_2)] \\ \sin\phi(\mathbf{V}_2) & \end{bmatrix} \exp[j\theta_{e_2}(\nu_2)]$$

Since the subspaces $\mathbf{V}_1$ and $\mathbf{V}_2$ must be orthogonal, it is easy to verify that $\sin\phi(\mathbf{V}_2) = \cos\phi(\mathbf{V}_1)$ and that $\exp[-j\psi(\mathbf{V}_2)] = -\exp[-j\psi(\mathbf{V}_1)]$. If, for notational brevity, we define $\phi_v \triangleq \phi(\mathbf{V}_1)$ and $\psi_v \triangleq \psi(\mathbf{V}_1)$, then the left singular vectors have the parameterization

$$\mathbf{V} = \begin{bmatrix} \cos\phi_v & -\sin\phi_v e^{-j\psi_v} \\ \sin\phi_v e^{j\psi_v} & \cos\phi_v \end{bmatrix} \cdot \begin{bmatrix} \exp[j\theta_{e_1}(\nu_1)] & 0 \\ 0 & \exp[j\theta_{e_2}(\nu_2)] \end{bmatrix} \tag{9.5.3}$$

An analogous parameterization of the right singular vectors yields

$$\mathbf{U} = \begin{bmatrix} \cos\phi_u & -\sin\phi_u e^{-j\psi_u} \\ \sin\phi_u e^{j\psi_u} & \cos\phi_u \end{bmatrix} \cdot \begin{bmatrix} \exp[j\theta_{e_1}(u_1)] & 0 \\ 0 & \exp[j\theta_{e_2}(u_2)] \end{bmatrix} \cdot \tag{9.5.4}$$

Recall that the phase of each singular vector is not well-defined, but that the phase difference (9.3.8)

$$\delta\theta_i(k) = \theta_{e_i}(\nu_k) - \theta_{e_i}(u_k) \tag{9.5.5}$$

between them is. Substituting (9.5.3)-(9.5.5) into (9.5.1) yields a parameterization of the singular value decomposition of a matrix $L \in \mathbb{C}^{2\times 2}$:

(9.5.6)

$$L = \begin{bmatrix} \cos\phi_v & -\sin\phi_v e^{-j\psi_v} \\ \sin\phi_v e^{j\psi_v} & \cos\phi_v \end{bmatrix} \cdot \begin{bmatrix} \sigma_1 e^{j\delta\theta_1(1)} & 0 \\ 0 & \sigma_2 e^{j\delta\theta_2(2)} \end{bmatrix} \cdot \begin{bmatrix} \cos\phi_u & -\sin\phi_u e^{-j\psi_u} \\ \sin\phi_u e^{j\psi_u} & \cos\phi_u \end{bmatrix}^H$$

Note that (9.5.6) contains 8 real parameters, as expected. This parameterization turns out to be quite useful in relating properties of the singular value decomposition to the elements of the matrix expressed in the standard basis.

For later reference, observe this notational shorthand implies that the phase-zero reference vectors in each singular subspace become

$$\rho_{e_1}(V_1) = \begin{bmatrix} \cos\phi_v \\ \sin\phi_v e^{j\psi_v} \end{bmatrix}, \quad \rho_{e_2}(V_2) = \begin{bmatrix} -\sin\phi_v e^{-j\psi_v} \\ \cos\phi_v \end{bmatrix} \tag{9.5.7}$$

and

$$\rho_{e_1}(U_1) = \begin{bmatrix} \cos\phi_u \\ \sin\phi_u e^{j\psi_u} \end{bmatrix}, \quad \rho_{e_2}(U_2) = \begin{bmatrix} -\sin\phi_u e^{-j\psi_u} \\ \cos\phi_u \end{bmatrix} . \tag{9.5.8}$$

9.6. Summary and Conclusions

In this chapter we have described the structure of the complex unit sphere in some detail. In particular, we have shown that a unit vector may be specified by giving one set of coordinates describing its direction and another coordinate analogous to scalar phase. These notions were made precise in Section 9.2 using the concept of a principal fiber bundle, specifically, a circle bundle over complex projective space. We have shown that it is not possible to obtain a single measure of phase applicable to all unit vectors in $\mathbb{C}^n$. However, several local measures of phase (local trivializations) may be obtained. To each of these, there corresponds a local rule for choosing phase-zero reference vectors (local sections). Transition functions may be used to fit these locally defined objects together on the overlap of their domains of definition.

In Section 9.3 these ideas were used to obtain measures of phase difference between a pair of singular vectors. Many such measures can be obtained, including the physically motivated one introduced in Chapter 6. Again, these measures of phase difference can all be related using δ-transition functions.

In Section 9.4 we discussed a method for assigning coordinates to complex projective space. We used these coordinates in Section 9.5 to obtain a parameterization of the transfer function of a system with two inputs and two outputs.

The results of this chapter show that the action of a linear system upon an input of the form $u(t) = ue^{s_0 t}$ can be decomposed into three components. One of these is a change in the *magnitude* of the signal. Another is a change in the *direction*, or subspace spanned, by the signal. These two components are well known. However, as an argument based upon the degrees of freedom needed to specify a signal will show, they fail to completely describe the action of the linear system. Indeed, the results of this chapter show that a linear system changes one additional property, analogous to the *phase*, of a vector-valued signal. Although this is perhaps intuitively obvious, it nonetheless takes a careful description of the fiber bundle structure to make this notion precise. In the next two chapters, we shall see that each singular value gain is related to the phase difference between the associated pair of singular vectors by a set of equations analogous to the scalar Cauchy-Riemann equations and Bode integral. The relevant difference between these equations and their scalar counterparts is that additional terms appear due to the motion of the singular subspaces. The results of this chapter will prove necessary both to derive these equations as well as to obtain insight into their properties.

CHAPTER 10

DIFFERENTIAL EQUATIONS FOR SINGULAR VECTORS AND FAILURE OF THE CAUCHY-RIEMANN EQUATIONS

10.1. Introduction

We have argued, in Chapter 8, that each singular value of a matrix transfer function should contain information about some other property of the matrix. Since the logarithm of a singular value is not harmonic, however, this additional information cannot take the form of a direct analogue to scalar phase. We also showed that the nonexistence of an analogue to scalar phase is due to the motion of the singular subspaces with frequency. To study this behavior in greater detail, in Chapter 9 we introduced a precise way to describe each singular vector by specifying its direction and its phase. The purpose of the present chapter is to derive differential equations describing how the singular vectors of a matrix transfer function change with frequency. By using the geometric structure introduced in Chapter 9 to study these equations, we shall be able to answer several of the questions posed in Chapter 8.

The remainder of this chapter is organized as follows. In Section 10.2 we develop formulas for the derivatives of the singular vectors, and use these formulas to compute the Laplacian of the logarithm of a singular value. As expected, this Laplacian is generally nonzero. Some additional geometric concepts are introduced in Section 10.3; specifically, that of a *connection* in a fiber bundle. The structure associated to this connection allows us to make the following statements precise. Given a pair of singular vectors corresponding to a distinct singular value, the motion of these vectors may be decomposed into a change in the phase difference between them as well as changes in the directions in which they lie. Furthermore, each singular value "gain" is related to the phase difference between the associated pair of singular vectors via equations analogous to the Cauchy-Riemann equations. The key difference is the presence of additional terms, due to the motion of the singular subspaces. In Section 10.4 we present an alternate method for deriving the differential equations which must be satisfied by a matrix transfer function after a change of coordinates. Section 10.5 contains a summary and conclusions.

10.2. Derivatives of Singular Vectors

Consider the singular value decomposition of $M(s)$, a matrix transfer function. Generally, the singular values and vectors of $M(s)$ will vary with the complex frequency variable $s = x + jy$. Assuming that the singular values of $M(s)$ are distinct at a point of the complex plane, they must satisfy the partial differential equations (8.3.4) at that point. In this section we shall derive partial differential equations which must be satisfied by the singular *vectors*. In so doing, we shall re-encounter the differential form (8.3.6) and use it to discuss a relation between each singular value and the associated pair of singular vectors.

Recall from Chapter 9 that $2n$ real numbers are needed to describe a vector in $\mathbb{C}^n$. Of these, $2n-2$ are needed to describe the subspace in which the vector lies, one is needed to specify the magnitude of the vector, and one is needed to describe the "phase" of the vector. Given a pair of singular vectors associated with a distinct singular value, the subspaces spanned by these vectors are well-determined, as is their magnitude. Although the phase of each vector is not well-determined, the phase difference between them is, and is governed by the equation $\text{Im}[v^H M u] = 0$. From this equation, it is clear that if (u,v) is a valid pair of singular vectors, then so is $e^{ja}(u,v)$, $\forall \ a \in I\!R$. We shall see that this degree of freedom manifests itself in the differential equations which a pair of singular vectors must satisfy.

It is not *a priori* obvious that the singular vectors are partially differentiable with respect to x and y. Indeed it is clear that this property can hold only if the degree of freedom available to choose these vectors is used appropriately. It turns out that there exist infinitely many ways to choose a pair of singular vectors so that the desired partial derivatives exist.[30] Indeed, a procedure for constructing a set of singular vectors with the desired properties is given in [MaH83]. An alternate procedure, which uses results from[Kat82], is presented in[Fre85]. The latter reference also contains a comparison of the two procedures. For purposes of brevity, we shall only present a *formal* derivation of the formula for a differential of a singular vector, omitting the tedious details needed to prove rigorously that the desired differentials exist. We shall also consider only the case that the singular values of $M(s)$ are distinct. Extensions to the case of multiple singular values are found in [FLC82] and[Fre85]. It will prove convenient to state our results using differential notation (Appendix B).

Theorem 10.2.1: Let $M(s)$ take values in $\mathbb{C}^{n \times n}$ and suppose that $M(s)$ is analytic, nonsingular and has distinct singular values at a point $s_0 \in \mathbb{C}$. Then there exists a neighborhood $B(s_0)$ so that these properties hold $\forall \ s \in B(s_0)$. Associated to each singular value $\sigma_i(s)$ there exists a pair of left and right singular vectors, denoted $v_i(s)$ and $u_i(s)$, respectively, which have the property that the differentials dv_i and du_i exist in $B(s_0)$. These differentials must satisfy the following equations:

$$\begin{aligned} du_i &= \sum_{k=1}^{n} u_k u_k^H du_i \\ dv_i &= \sum_{k=1}^{n} v_k v_k^H dv_i \end{aligned} \qquad (10.2.1)$$

[30] There is generally *no* way to choose the singular vectors so that they are differentiable with respect to the complex variable s. Hence the singular vectors, like the singular values, are not complex analytic.

where, for $k \neq i$,

$$u_k^H du_i = \left(\frac{1}{\sigma_i^2-\sigma_k^2}\right)\left[\sigma_i u_k^H dM^H v_i + \sigma_k v_k^H dMu_i\right]$$
$$v_k^H dv_i = \left(\frac{1}{\sigma_i^2-\sigma_k^2}\right)\left[\sigma_k u_k^H dM^H v_i + \sigma_i v_k^H dMu_i\right] \quad . \tag{10.2.2}$$

The components $u_i^H du_i$ and $v_i^H dv_i$ are arbitrary subject to the constraints

$$\mathrm{Re}[u_i^H du_i] = \mathrm{Re}[v_i^H dv_i] = 0 \tag{10.2.3}$$

and

$$v_i^H dv_i - u_i^H du_i = \frac{j}{\sigma_i} \mathrm{Im}[v_i^H dMu_i] \tag{10.2.4}$$

■

Formal Proof: Existence of the set $B(s_0)$ and the differentials dv_i and du_i are discussed in[Fre85]. Using various properties of singular values and vectors, we have, for $k \neq i$,

$$M^H Mu_i = \sigma_i^2 u_i$$

$$\Rightarrow \quad d(M^H M)u_i + M^H Mdu_i = 2\sigma_i u_i d\sigma_i + \sigma_i^2 du_i$$

$$\Rightarrow \quad u_k^H d(M^H M)u_i + \sigma_k^2 u_k^H du_i = 0 + \sigma_i^2 u_k^H du_i$$

$$\Rightarrow \quad u_k^H du_i = \left(\frac{1}{\sigma_i^2-\sigma_k^2}\right)\left[u_k^H dM^H Mu_i + u_k^H M^H dMu_i\right]$$

from which the result follows. A similar computation yields $v_k^H dv_i$.

To show (10.2.3) note that

$$u_i^H u_i \equiv 1$$

$$\Rightarrow \quad du_i^H u_i + u_i^H du_i = 0$$

$\Rightarrow \qquad 2\text{Re}[u_i^H du_i] = 0$

Finally, we have

$$v_i^H M u_i = \sigma_i$$

$\Rightarrow \qquad \sigma_i (dv_i^H v_i + u_i^H du_i) + v_i^H dM u_i = d\sigma_i$

Using (8.3.4)

$\Rightarrow \qquad \sigma_i (dv_i^H v_i + u_i^H du_i) + j\,\text{Im}[v_i^H dM u_i] = 0$

from which the result follows.

■

Recalling that each singular vector may be described by specifying its *direction* and its *phase* yields an interesting interpretation of Theorem 10.2.1. The fact that each singular subspace, or direction, is well-defined as an eigenspace of MM^H or $M^H M$ allows its differential to be calculated straightforwardly. Indeed, the differential of each subspace is uniquely determined by the appropriate set of $n-1$ differential forms $v_k^H dv_i$ or $u_k^H du_i$, $k \neq i$. It seems reasonable that the remaining forms $v_i^H dv_i$ and $u_i^H du_i$ determine the differential changes in the phase of each singular vector. That these forms are arbitrary subject only to a constraint on their difference (10.2.4) is consistent with the fact that only the phase difference between each pair of singular vectors is uniquely determined.

Equation (10.2.4) may be used to answer a question posed at the close of Section 8.3. There we argued that the analyticity of $M(s)$ implies that each one-form $*d\log\sigma_i$ contains information about the differential of $M(s)$. By combining (8.3.6) and (10.2.4), we can show that each of these forms determines a property of the associated pair of singular vectors.

Corollary 10.2.2 Under the assumptions of Theorem 10.2.1, each singular value and associated pair of singular vectors must satisfy

$$*d\log\sigma_i = \frac{1}{j}\,[v_i^H dv_i - u_i^H du_i] \tag{10.2.5}$$

■

Proof:

$$\begin{aligned} v_i^H dv_i - u_i^H du_i &= \frac{j}{\sigma_i}\,\mathrm{Im}[v_i^H dM u_i] \\ &= \frac{j}{\sigma_i}\,\mathrm{Im}[v_i^H \frac{\partial M}{\partial x} u_i]dx + \frac{j}{\sigma_i}\,\mathrm{Im}[v_i^H \frac{\partial M}{\partial y} u_i]dy \\ &= j[-\frac{\partial\log\sigma_i}{\partial y}dx + \frac{\partial\log\sigma_i}{\partial x}dy] \quad . \end{aligned}$$

■

In view of the discussion following Theorem 10.2.1, one might be tempted to conjecture that (10.2.5) relates each singular value to the phase difference between the associated pair of singular vectors. However, in Chapter 8, we showed by counterexample that (10.2.5) cannot be *equal* to the differential of the phase difference; the singular values of (8.3.1) have logarithms whose Laplacians are nonzero. Using (10.2.5), we can readily obtain a *general* formula for the Laplacian of the logarithm of a singular value.

Theorem 10.2.3: Under the assumptions of Theorem 10.2.1, the Laplacian of the logarithm of each singular value must satisfy

$$\begin{aligned} \nabla^2\log\sigma_i = \sum_{k\neq i}\left[\frac{1}{\sigma_i^2-\sigma_k^2}\right]\Big\{ & |v_k^H\frac{\partial M}{\partial x}u_i|^2 + |v_k^H\frac{\partial M}{\partial y}u_i|^2 \\ & + |v_i^H\frac{\partial M}{\partial x}u_k|^2 + |v_i^H\frac{\partial M}{\partial y}u_k|^2 \Big\} \end{aligned} \tag{10.2.6}$$

■

Proof: See Appendix D. ∎

Theorem 10.2.3 proves that, except in special cases, $\log\sigma_i$ cannot be a harmonic function even *locally*. One implication of this fact is that $\log\sigma_i$ can have no harmonic conjugate. Another implication is that the functions $\log\sigma_i$ can have local maxima and/or local minima (properties which are prohibited by the well-known mean value property of harmonic functions[Con78]). In particular, the fact that $\nabla^2\log\overline{\sigma} \geq 0$ may be used to show [FaK80] that $\log\overline{\sigma}$ is *subharmonic* and can posses local minima but no local maxima. Similarly, the fact that $\nabla^2\log\underline{\sigma} \leq 0$ implies that $\log\underline{\sigma}$ is *superharmonic* and can possess local maxima but no local minima. A number of useful design implications of the former fact have been presented by Boyd and Desoer[BoD84], who prove that $\log\overline{\sigma}$ is subharmonic even in regions where $\overline{\sigma}$ can have multiplicity greater than one.

Useful insights into the behavior of scalar harmonic and analytic functions in general (and transfer functions in particular) may be gained from an analogy with two-dimensional potential theory, e.g.,[Gui49]. It is interesting to interpret the fact that $\nabla^2\log\sigma_i \neq 0$ in this context.

Identify $\mathbb{C}$ the complex plane, with $\mathbb{R}^2$, and think of the gradient vector

$$\nabla\log\sigma_i = \left[\frac{\partial\log\sigma_i}{\partial x}, \frac{\partial\log\sigma_i}{\partial y}\right] \tag{10.2.7}$$

as being the velocity vector field of a fluid flowing in steady state along $\mathbb{R}^2$. Let $C \in \mathbb{R}^2$ be a simple closed curve with interior D, and suppose that C is parameterized by $\gamma(t) = (x(t), y(t))$, $a \leq t \leq b$. Assume further that γ is traversed counterclockwise with unit speed: $\| \frac{d\gamma}{dt} \|_2 \equiv 1$. Hence the outward pointing unit normal to $\gamma(t)$ is given by $N \triangleq \left[\frac{dy}{dt}, -\frac{dx}{dt}\right]$.

Using these ideas, it makes sense to ask whether the net fluid flowing into the region bounded by C is positive, negative, or zero. Clearly, if

$$\nabla \log\sigma_i \cdot N = \left[- \frac{\partial \log\sigma_i}{\partial y} \frac{dx}{dt} + \frac{\partial \log\sigma_i}{\partial x} \frac{dy}{dt} \right] \tag{10.2.8}$$

is positive at a point of the boundary, then fluid is flowing out of the region D at that point. The net fluid flow through D is given by the integral

$$\begin{aligned} \text{net flow} &= \int_C \left[\frac{\partial \log\sigma_i}{\partial x_1}, \frac{\partial \log\sigma_i}{\partial y} \right] \cdot (dy\ ,\ -dx) \\ &= \int_C \left[- \frac{\partial \log\sigma_i}{\partial y} dx + \frac{\partial \log\sigma_i}{\partial x} dy \right] \\ &= \int_a^b \left[- \frac{\partial \log\sigma_i}{\partial y} \frac{dx}{dt} + \frac{\partial \log\sigma_i}{\partial x} \frac{dy}{dt} \right] dt \end{aligned} \tag{10.2.9}$$

Using Stokes' Theorem (Appendix B; also[Spi65]) the net flow is given by the surface integral

$$\text{net flow} = \int_D \left[\frac{\partial^2 \log\sigma_i}{\partial x^2} + \frac{\partial^2 \log\sigma_i}{\partial y^2} \right] dx \wedge dy \ . \tag{10.2.10}$$

From (10.2.10), it follows that if $\log\sigma_i$ were harmonic, then the net fluid flow would be zero. The fact that $\nabla^2 \log\bar{\sigma} \geq 0$ implies that the net flow of the associated fluid through D is *positive*. The flow behaves as though *sources* of fluid were distributed continuously throughout the region D. On the other hand, the fact that $\nabla^2 \log\underline{\sigma} \leq 0$ implies that the associated net flow is *negative*. In this case, the fluid behaves as though *sinks* were distributed continuously throughout D. For intermediate singular values, no general statements can be made. However, since $\det M(s)$ is an analytic function, it follows that $\log|\det M| = \prod_{i=1}^{n} \log\sigma_i$ is harmonic. Hence the sum of the net flows (10.2.10) over all singular values must be equal to zero. In particular, when $n = 2$, it is as though the excess fluid entering the region D from the flow associated with $\nabla \log\underline{\sigma}$ is somehow transferred and leaves the region in the flow associated with

$\nabla \log \bar{\sigma}$.

To summarize, each singular value is related to a property of the associated singular vector pair via the differential form (10.2.5). Although it seems plausible that this form is related to the phase difference between the pair of singular vectors, Theorem 10.2.3 shows conclusively that (10.2.5) cannot *equal* the differential of the phase difference. Nevertheless, we shall see in the next section that (10.2.5) may be used to obtain equations which show precisely how each singular value is related to the phase difference between its pair of singular vectors.

10.3. Connection Forms and Generalized Cauchy-Riemann Equations

The purpose of this section is to use the geometric framework developed in Chapter 9 to study the relation between a singular value and its associated pair of singular vectors. Specifically, we shall use local trivializations to decompose differential changes in a pair of singular vectors into differential changes in the subspaces which they span and a differential change in the phase difference between them. It will follow that each singular value is related to the phase difference between its pair of singular vectors by a set of partial differential equations resembling the Cauchy-Riemann equations. The key difference is that additional terms, due to the singular subspaces, appear in the equations. These additional terms each satisfy the definition of a *connection* in a fiber bundle. They measure how the motion of the singular subspaces with frequency prevents the singular value and the phase difference between singular vectors from satisfying the well-known relations between scalar gain and phase. Before proceeding, we need to review some additional ideas arising in the study of principal fiber bundles.

Recall that each principal fiber bundle consists, in part, of a Lie group G. It turns out that each Lie group has an associated *Lie algebra*, denoted $\tilde{G}$, which describes the behavior of the group in the neighborhood of its identity element[Ble81]. Let the tangent space to the identity element of G be denoted $T_e G$. For our purposes, the Lie algebra associated with G is just $T_e G$ together with a certain algebraic structure[Ble81]. The Lie group in which we are interested is the scalar unitary group $Un(1) = \{e^{ja} : a \in IR\}$. To determine the Lie algebra associated with $Un(1)$, consider a smooth curve $a(t) \in IR$ with $a(0) = 0$. There exists a corresponding curve $\exp(ja(t)) \in Un(1)$ which passes through the identity at $t = 0$. The derivative of this curve at $t = 0$ is $\frac{d}{dt}\exp(ja(t))\Big|_{t=0} = j\frac{da}{dt}\Big|_{t=0}$. Hence the Lie algebra of $Un(1)$, identified with the tangent space to the identity, is $\tilde{Un}(1) = T_1 Un(1) = \{jb : b \in IR\}$. Note that the tangent space at a general point $u \in Un(1)$ is the set $e^{ja}\tilde{Un}(1) \triangleq \{e^{ja} \cdot jb : b \in IR\}$. Thus there exists a 1-1 correspondence between the tangent space at any point in $Un(1)$ and the tangent space at the identity.

Definition 10.3.1[Ble81]: A *connection* assigns to each local trivialization $T_\alpha : \pi^{-1}(N_\alpha) \rightarrow N_\alpha \times G$ a $\tilde{G}$-valued one-form ω_α on N_α. If T_β is another local trivialization and $g_{\alpha\beta}: N_\alpha \cap N_\beta \rightarrow G$ is the transition function from T_α to T_β, then it is required that

$$\omega_\beta = g_{\alpha\beta}^{-1} dg_{\alpha\beta} + g_{\alpha\beta}^{-1}\omega_\alpha g_{\alpha\beta} \tag{10.3.1}$$

■

The fact that ω_α is a $\tilde{G}$-valued one-form on N_α means that ω_α is a linear operator mapping each vector tangent to a point of N_α to an element of the Lie algebra $\tilde{G}$. Although connection forms corresponding to different local trivializations generally yield different values on the overlap of their domains of definition, these values may be related in a straightforward fashion using the transition functions.

To obtain the connection forms of interest to us, consider a local trivialization, or measure of phase, defined as in (9.2.5)

$$\begin{aligned} T_w &: \pi^{-1}(N_w) \to N_w \times S^1 \\ T_w(z) &= ([z], \exp[j\theta_w(z)]) \quad . \end{aligned} \tag{10.3.2}$$

The corresponding local section, or choice of phase-zero reference vector, is given by

$$\begin{aligned} \rho_w &: N_w \to S^{2n-1} \\ \rho_w(\mathbf{Z}) &= T_w^{-1}(\mathbf{Z},1) \end{aligned} \tag{10.3.3}$$

Theorem 10.3.2 Consider the set of local trivializations defined by (10.3.2) and the associated local sections (10.3.3). The rule which assigns to each local trivialization the one-form

$$\omega_w \triangleq \rho_w^H \, d\rho_w \tag{10.3.4}$$

satisfies the definition of a connection in the circle bundle over $\mathbb{C}P^{n-1}$.

■

Proof: First, we need to show that ω_w takes values in the Lie algebra $\tilde{Un}(1)$. To do this, let $\mathbf{Z}:[-1,1] \to N_w$ be a smooth curve parameterized by t. Denote the tangent vector to the curve by $\frac{d\mathbf{Z}}{dt}$, and consider the composite map $\rho_w \circ \mathbf{Z}:[-1,1] \to S^{2n-1}$. Evaluating (10.3.4) on $\frac{d\mathbf{Z}}{dt}$ yields

$$\begin{aligned} \omega_w\left(\frac{d\mathbf{Z}}{dt}\right) &= \rho_w^H d\rho_w\left(\frac{d\mathbf{Z}}{dt}\right) \\ &= \rho_w^H \frac{d(\rho_w \circ \mathbf{Z})}{dt} \quad . \end{aligned} \tag{10.3.5}$$

Since $\rho_w(\mathbf{Z})$ is, by definition, a unit vector, it follows easily that $\omega_w(\frac{d\mathbf{Z}}{dt}) \in \tilde{Un}(1)$. Furthermore, since local sections corresponding to different local trivializations are related, for $\mathbf{Z} \in N_{w_i} \cap N_{w_k}$, by

$$\rho_{w_k}(\mathbf{Z}) = \rho_{w_i}(\mathbf{Z}) g_{ik}(\mathbf{Z}) \tag{10.3.6}$$

and since $g_{ik}(\mathbf{Z}) \in Un(1)$, it follows that

$$d\rho_{w_k} = d\rho_{w_i}\, g_{ik} + \rho_{w_i}\, dg_{ik}$$

$$\begin{aligned} \Rightarrow \qquad \rho_{w_k}^H d\rho_{w_k} &= \rho_{w_k}^H d\rho_{w_i} + \rho_{w_k}^H \rho_{w_i} dg_{ik} \\ &= \bar{g}_{ik}\, \rho_{w_i}^H d\rho_{w_i}\, g_{ik} + \bar{g}_{ik}\, dg_{ik} \\ &= g_{ik}^{-1}\, dg_{ik} + \rho_{w_i}^H\, d\rho_{w_i} \quad . \end{aligned} \tag{10.3.7}$$

Hence the one-forms defined by (10.3.4) satisfy the condition (10.3.1) and the rule assigning these one-forms to each local trivialization is indeed a connection.

■

We shall see shortly that the connection forms (10.3.4) arise naturally in studying the phase difference between a pair of singular vectors. First, we need to introduce a local trivialization with which to measure phase difference.

Consider a matrix transfer function $M(s)$ satisfying the conditions stated in Theorem 10.2.1. Let σ, v, and u denote a singular value and associated pair of left and right singular vectors.[31] By identifying $\mathbf{V}$ and $\mathbf{U}$, the associated pair of left and right singular subspaces, with the corresponding points of projective space, we may obtain maps $\mathbf{V} : B(s_0) \to \mathbb{C}P^{n-1}$ and $\mathbf{U} : B(s_0) \to \mathbb{C}P^{n-1}$. It is easy to see that there always exists a local trivialization (9.2.5) whose domain of definition includes both $\mathbf{V}(s_0)$ and $\mathbf{U}(s_0)$. (Any local trivialization defined using a vector w with $w^H v \neq 0$ and $w^H u \neq 0$ will work). Let this local trivialization be denoted $T_w : \pi^{-1}(N_w) \to N_w \times S^1$. By continuity, this local trivialization also contains within its domain of definition the values of $\mathbf{V}(s)$ and $\mathbf{U}(s)$ for all points in some neighborhood

[31]We drop the subscript ordering the singular values for notational simplicity.

$B_w(s_0) \subseteq B(s_0)$. (Note that the size of this neighborhood will generally depend upon the choice of local trivialization T_w). We summarize this discussion in the following lemma.

Lemma 10.3.3: Consider a matrix transfer function $M(s)$ satisfying the conditions stated in Theorem 10.2.1. Let σ denote one of the singular values of $M(s)$ and let v, u, V, and U denote the associated left and right singular vectors and subspaces. There exists a local trivialization $T_w : \pi^{-1}(N_w) \rightarrow N_w \times S^1$ and a neighborhood $B_w(s_0) \subseteq B(s_0)$ such that the left and right singular vectors may be expressed as

$$v(s) = \rho_w(\mathrm{V}(s))\exp[j\theta_w(v(s))]$$

and (10.3.8)

$$u(s) = \rho_w(\mathrm{U}(s))\exp[j\theta_w(u(s))]$$

for all $s \in B_w(s_0)$.

■

The unit vectors $\rho_w(\mathrm{V}(s))$ and $\rho_w(\mathrm{U}(s))$ in (10.3.8) are the values of the composite maps $\rho_w \circ \mathrm{V}$ and $\rho_w \circ \mathrm{U}$ at the point s, where ρ_w is the local section corresponding to the local trivialization T_w. These vectors satisfy the relations

$$\begin{aligned} &\rho_w \circ \mathrm{V} : B_w(s_0) \rightarrow S^{2n-1} \\ &\rho_w(\mathrm{V}(s)) = T_w^{-1}(\mathrm{V}(s),1) \end{aligned} \tag{10.3.9}$$

and

$$\begin{aligned} &\rho_w \circ \mathrm{U} : B_w(s_0) \rightarrow S^{2n-1} \\ &\rho_w(\mathrm{U}(s)) = T_w^{-1}(\mathrm{U}(s),1) \quad . \end{aligned} \tag{10.3.10}$$

It is instructive to illustrate the map $\rho_w \circ \mathrm{V}$ using Figure 10.3.1. A similar diagram may be constructed for $\rho_w \circ \mathrm{U}$. Figure 10.3.1 highlights the fact that the vector $\rho_w(\mathrm{V}(s))$ in (10.3.8)

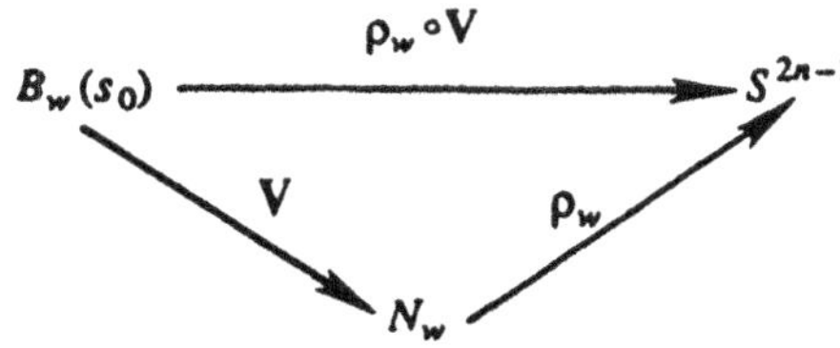

Figure 10.3.1: Factorization of the Map $\rho_w \circ V$

may be obtained by first computing the left singular subspace corresponding to the singular value σ and then using the local section ρ_w to compute the "phase zero" reference vector in that subspace. This factorization will prove useful in our study of how the form (10.2.5) is related to the phase difference between v and u.

Equations (10.3.8)-(10.3.10) show how local trivializations and sections may be used to describe each singular vector in terms of direction and phase. These equations may also be used to obtain expressions for the differential of the phase of each vector. Clearly, we must first insure that these differentials exist. From the discussion of Theorem 10.2.1 it follows that the subspaces V and U each have continuous partial derivatives with respect to x and y, and it is not hard to show that the sections $\rho_w(V)$ and $\rho_w(U)$ also have this property. Furthermore, the vectors $v(\cdot)$ and $u(\cdot)$ may be *chosen* so that the composite maps $\exp[j(\theta_w \circ v)]$: $B_w(s_0) \to S^1$ and $\exp[j(\theta_w \circ u)]$: $B_w(s_0) \to S^1$ each have continuous partial derivatives. Hence we may indeed compute the differentials dv and du $\forall\ s \in B_w(s_0)$. A detailed discussion of these differentiability properties using results from [Kat82] is found in[Fre85].

Theorem 10.3.4 Consider the pair of singular vectors (10.3.8)-(10.3.10), and assume that these have been chosen so that $\exp[j(\theta_w \circ v)]$ and $\exp[j(\theta_w \circ u)]$ have continuous partial derivatives in $B_w(s_0)$. Then, $\forall\ s \in B_w(s_0)$, the following relations must be satisfied:

$$v^H dv = \rho_w^H d(\rho_w \circ V) + jd(\theta_w \circ v)$$
$$u^H du = \rho_w^H d(\rho_w \circ U) + jd(\theta_w \circ u) \quad . \qquad (10.3.11)$$

■

Proof:

From (10.3.8) it follows that

$$dv = d(\rho_w \circ V)\exp[j(\theta_w \circ v)] + vjd(\theta_w \circ v)$$

and (10.3.12)

$$du = d(\rho_w \circ U)\exp[j(\theta_w \circ u)] + ujd(\theta_w \circ u) \quad .$$

The result follows by substituting (10.3.8) into the expressions for $v^H dv$ and $u^H du$.

■

Since the phase of each singular vector is not well-defined, one might conclude that equations (10.3.11) have no physical significance. Indeed, the importance of these equations lies primarily in the fact that they can be used to compute the *phase difference* (9.3.1) between the pair of vectors. The value of phase difference is obtained from the following composite mapping $\delta\theta_w(u,v)$:

$$\exp[j\delta\theta_w \circ (u,v)] : B_w(s_0) \rightarrow S^1$$
$$\exp[j\delta\theta_w(u(s),v(s))] = \exp[j\theta_w(v(s))] \exp[-j\theta_w(u(s))] \quad . \qquad (10.3.13)$$

Not only is the phase difference well-defined, but it is straightforward to show [Fre85] that it has continuous partial derivatives. Thus the differential of $\delta\theta_w \circ (u,v)$ exists and can be obtained as a corollary to Theorem 10.3.4.

Corollary 10.3.5: The phase difference (10.3.13) between the pair of singular vectors (10.3.8) satisfies

$$v^H dv - u^H du = \rho_w^H d(\rho_w \circ V) - \rho_w^H d(\rho_w \circ U) + jd(\delta\theta_w \circ (u,v)) \qquad (10.3.14)$$

at each point of $B_w(s_0)$.

■

Equation (10.3.14) shows precisely how the difference between the components $v^H dv$ and $u^H du$ is related to the differential of the phase difference between u and v. To understand the nature of this relation, it is necessary to first understand the terms $\rho_w^H d(\rho_w \circ V)$ and $\rho_w^H d(\rho_w \circ U)$ which appear in (10.3.14). The simplest way to do this is to first analyze the individual equations (10.3.11) and then combine the results to draw conclusions about (10.3.14). We shall proceed by discussing the first of equations (10.3.11), governing the behavior of the left singular vector. A similar analysis applies to the right singular vector.

First, recall the discussion following Theorem 10.2.1, where we argued that the differential form $v^H dv$ should be related to the differential of the phase of the left singular vector. Theorem (10.3.4) shows that the values of these differentials are not, in general, equal. The difference between them is determined by the value of the connection form $\rho_w^H d\rho_w$ evaluated on the tangent vector to the path through $\mathbb{C}P^{n-1}$ traversed by the left singular subspace. Some insight into this phenomenon can be obtained by showing that the factorization of the section map $\rho_w \circ V$ (Figure 10.3.1) induces a factorization of the connection map $\rho_w^H d(\rho_w \circ V)$.

By assumption, the image of the subset $B_w(s_0) \subseteq \mathbb{C}$ under the left singular subspace map V lies in the subset $N_w \subseteq \mathbb{C}P^{n-1}$; i.e., $V(B_w(s_0)) \subseteq N_w \subseteq \mathbb{C}P^{n-1}$. Let the tangent spaces to $B_w(s_0)$ and N_w be denoted $TB_w(s_0)$ and TN_w, respectively, and consider a smooth curve $\gamma(t) \in B_w(s_0)$ with tangent vector $\frac{d\gamma}{dt} = \frac{dx}{dt} + j\frac{dy}{dt}$. Then dV, the differential of the left singular subspace, may be viewed as a mapping from $TB_w(s_0)$ to TN_w:

$$dV : TB_w(s_0) \rightarrow TN_w$$
$$dV\left(\frac{d\gamma}{dt}\right) = \frac{d(V \circ \gamma)}{dt} \quad . \tag{10.3.15}$$

Since $V(t) = (V \circ \gamma)(t)$ is a curve through $\mathbb{C}P^{n-1}$ with tangent vector $\frac{dV}{dt}$, the connection form $\rho_w^H d\rho_w$ may be evaluated along the tangent vector $\frac{dV}{dt}$ to obtain an element of the Lie algebra $\bar{U}n(1)$:

$$\rho_w^H d\rho_w \left[\frac{dV}{dt}\right] = \rho_w^H \frac{d(\rho_w \circ V)}{dt} \quad . \tag{10.3.16}$$

The value of the differential form $\rho_w^H d(\rho_w \circ V)$ evaluated on $\frac{d\gamma}{dt}$ is then obtained from the composition of (10.3.15) and (10.3.16):

$$\rho_w^H d(\rho_w \circ V) \left[\frac{d\gamma}{dt}\right] = \rho_w^H \frac{d(\rho_w \circ V \circ \gamma)}{dt} \quad . \tag{10.3.17}$$

Equations (10.3.15)-(10.3.17) are illustrated in Figure 10.3.2: In Figure 10.3.2, the connection form $\rho_w^H d\rho_w$ is defined in the fiber bundle S^1 over $\mathbb{C}P^{n-1}$. The differential form $\rho_w^H d(\rho_w \circ V)$, on the other hand, is a connection form defined in the *pullback* [Ble81] of this bundle along the left singular subspace map V: $B_w(s_0) \rightarrow N_w \subseteq \mathbb{C}P^{n-1}$.

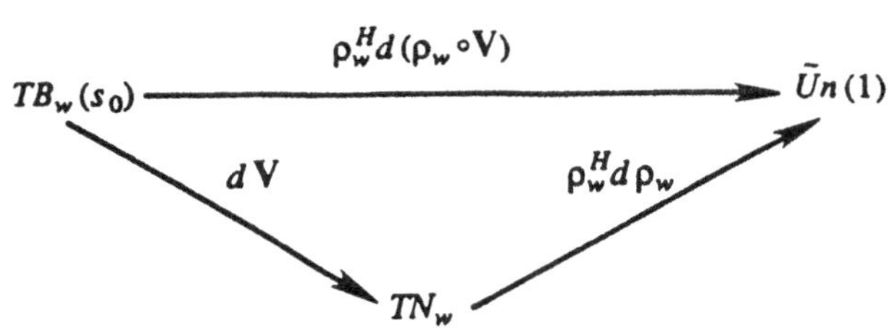

Figure 10.3.2. Factorization of the Connection Mapping.

To summarize, evaluating (10.3.11) on a tangent vector to a curve $\gamma \in \mathcal{C}$ shows that a discrepancy exists between the values of $v^H \frac{dv}{dt}\Big|_\gamma$ and $j\frac{d(\theta_w \circ v)}{dt}\Big|_\gamma$. This discrepancy is equal to $\rho_w^H \frac{d(\rho_w \circ V)}{dt}\Big|_\gamma$ and, as Figure 10.3.2 illustrates, its value is a function of the path through projective space traversed by the left singular subspace. A completely analogous discussion applies to the right singular vector and subspace. The differential form (10.3.14) therefore shows that the *difference* between the paths traversed by the left and right singular subspaces causes the value of $\frac{d(\delta\theta_w \circ (u,v))}{dt}\Big|_\gamma$ to deviate from that of $[v^H \frac{dv}{dt} - u^H \frac{du}{dt}]\Big|_\gamma$. Since the difference between these paths is responsible for the failure of $\log\sigma$ and $\delta\theta_w(u,v)$ to satisfy the Cauchy-Riemann equations and, ultimately, the gain-phase relations, this is a crucial observation.

Although $\log\sigma$ and $\delta\theta_w(u,v)$ fail to satisfy the usual classical results on analytic functions, certain "generalized" versions of these results do hold. Each of these generalized results is similar to the corresponding classical result, with the addition of a "correction term" depending upon the motion of the singular subspaces. The following corollary shows that Corollaries 10.2.2 and 10.3.4 may be combined to yield a generalized version of the Cauchy-Riemann equations.

Corollary 10.3.6: Assume that the conditions stated in Theorem 10.2.1 and Lemma 10.3.3 are satisfied. Then each singular value and its associated measure of phase difference must satisfy

$$*d\log\sigma = d(\delta\theta_w \circ (u,v)) + \frac{1}{j}[\rho_w^H d(\rho_w \circ V) - \rho_w^H d(\rho_w \circ U)] \qquad (10.3.18)$$

at each point of $B_w(s_0)$. Furthermore, (10.3.18) implies that $\log\sigma$ and $\delta\theta_w(u,v)$ must satisfy the following partial differential equations:

$$\frac{\partial \log\sigma}{\partial x} = \frac{\partial[\delta\theta_w(u,v)]}{\partial y} + \frac{1}{j}\left[\rho_w^H \frac{\partial[\rho_w(V)]}{\partial y} - \rho_w^H \frac{\partial[\rho_w(U)]}{\partial y}\right]$$
$$\frac{\partial \log\sigma}{\partial y} = -\frac{\partial[\delta\theta_w(u,v)]}{\partial x} - \frac{1}{j}\left[\rho_w^H \frac{\partial[\rho_w(V)]}{\partial x} - \rho_w^H \frac{\partial[\rho_w(U)]}{\partial x}\right] . \tag{10.3.19}$$

■

Proof: The differential form (10.3.18) follows from (10.2.5) and (10.3.14). Equations (10.3.19) follow by noting that

$$\begin{aligned} *d\log\sigma &= -\frac{\partial \log\sigma}{\partial y}dx + \frac{\partial \log\sigma}{\partial x}dy \\ &= \frac{\partial[\delta\theta(u,v)]}{\partial x}dx + \frac{1}{j}\left[\rho_w^H \frac{\partial[\rho_w(V)]}{\partial x} - \rho_w^H \frac{\partial[\rho_w(U)]}{\partial x}\right]dx \\ &+ \frac{\partial[\delta\theta(u,v)]}{\partial y}dy + \frac{1}{j}\left[\rho_w^H \frac{\partial[\rho_w(V)]}{\partial y} - \rho_w^H \frac{\partial[\rho_w(U)]}{\partial y}\right]dy . \end{aligned}$$

■

It is instructive to compare equations (10.3.19) with the Cauchy-Riemann equations (8.2.3). The latter constrain the gain and phase of a scalar transfer function, while the former express a relation between each singular value gain and the phase difference between the associated pair of singular vectors. It is easy to see that it is the motion of the singular subspaces with frequency which prevents the usual Cauchy-Riemann equations from being satisfied. This is consistent with our previous conclusion, in Section 8.4, that motion of the singular subspaces prevents $\log\sigma$ from being harmonic and hence $\sigma\exp[j\delta\theta_w(u,v)]$ from being complex analytic. The latter fact has several interpretations. For example, consider the conformal mapping property of analytic functions. This well-known property implies that, for a scalar transfer function, lines of constant gain and phase must intersect at right angles. Equations (10.3.19) show that lines of constant singular value "gain" and singular vector "phase difference" intersect at an angle determined by the partial derivatives of the singular subspaces.

The generalized Cauchy-Riemann equations developed in this section provide the foundation for our ultimate goal of deriving multivariable versions of the scalar gain-phase relations. The role they shall play for us is precisely the same as that played by the standard Cauchy-Riemann equations in the development of the classical scalar results.

10.4. Cauchy's Integral Theorem and an Alternate Method for Deriving Differential Equations

Motivated by the degrees-of-freedom argument in Section 8.1, we have derived a set of differential equations constraining the properties of a matrix transfer function. In particular, we have shown that knowledge of a singular value and the associated pair of singular subspaces suffices to determine the phase difference between the pair of singular vectors to within a constant. Our approach has been somewhat roundabout, in that we first developed differential equations for the singular values and vectors of $M(s)$ in terms of $\frac{\partial M}{\partial x}$ and $\frac{\partial M}{\partial y}$. We then used the fact that the elements of $M(s)$, in standard coordinates, must satisfy the Cauchy-Riemann equations to show that the singular values and vectors cannot be mutually independent functions of frequency.

As might be suggested by our degree-of-freedom argument, the generalized Cauchy-Riemann equations are not the only differential equations constraining the matrix transfer function. Indeed, it is straightforward to show, using Lemma 8.3.2 and Theorem 10.2.1, that the motion of the singular subspaces is subject to additioned constraints. If desired, these can be stated explicitly in terms of the latitude and longitude coordinates introduced in Section 9.4. Rather than take this approach, we shall instead present a systematic methodology for deriving the complete set of partial differential equations which must be satisfied by the singular values and vectors of a matrix transfer function.

The mathematical result we shall use is the well-known Cauchy Integral Theorem (e.g.,[Con78]).

Theorem 10.4.1. Let $f = g + jh$ be analytic on a simply connected region $G \subseteq \mathbb{C}$ and let C denote a simple closed curve in G. Then

$$\int_C f(s)ds = 0 \tag{10.4.1}$$

■

Let D denote the interior of C. Then Stokes' Theorem ([Spi65], pp.105-106) shows that

$$\int_C f ds = \int_D df \wedge ds$$

where

$$\begin{aligned} df \wedge ds &= \left[\frac{\partial f}{\partial x}\, dx + \frac{\partial f}{\partial y}\, dy\right] \wedge (dx + jdy) \\ &= j\left[\frac{\partial f}{\partial x} + j\frac{\partial f}{\partial y}\right] dx \wedge dy \quad . \end{aligned}$$

Hence, by (10.4.1), $\frac{\partial f}{\partial x} + j\frac{\partial f}{\partial y} = 0$; this is just an alternate statement of the Cauchy-Riemann equations (8.2.1). By contrast, the integral of the nonanalytic function $\log\sigma + j\,\delta\theta_w(u,v)$ must satisfy

$$\int_C [\log\sigma + j\,\delta\theta_w(u,v)]ds = \tag{10.4.2}$$

$$\int_D \left[\left[\frac{\partial \log\sigma}{\partial x} - \frac{\partial \delta\theta_w(u,v)}{\partial y}\right] + j\left[\frac{\partial \log\sigma}{\partial y} + \frac{\partial \delta\theta_w(u,v)}{\partial x}\right]\right] (dx \wedge dy)$$

which the generalized Cauchy-Riemann equations (10.3.19) show must be nonzero. Hence motion of the singular subspaces prevents the Cauchy Integral Theorem, as well as the Cauchy-Riemann equations, from being satisfied.

Let each element of the matrix transfer function $M(s)$ satisfy the assumptions of Theorem 10.4.1. Then $\int_C M ds = 0$. Equivalently,

$$dM \wedge ds = 0 \qquad \forall\ s \in D \quad . \tag{10.4.3}$$

Condition (10.4.3) is the key to seeing how the property of analyticity generalizes to nonstandard coordinates. To illustrate this, consider the singular value decomposition of M:

$$M = V\ \Sigma\ U^H \quad . \tag{10.4.4}$$

Then, from (10.4.3),

$$dM \wedge ds = (dV \wedge ds)\Sigma U^H + V(d\Sigma \wedge ds)U^H + V\Sigma(dU^H \wedge ds) = 0$$

$$\Rightarrow \qquad [dV\,\Sigma U^H + V d\Sigma U^H + V\Sigma dU^H] \wedge ds = 0$$

$$\Rightarrow \qquad \Sigma^{-1} d\Sigma \wedge ds = -\,[dU^H U + \Sigma^{-1} V^H dV\,\Sigma] \wedge ds \tag{10.4.5}$$

By expanding the differentials in (10.4.5) we may obtain $2n^2$ partial differential equations which must be satisfied by the singular values and vectors. It is easy to verify that the diagonal entries of (10.4.5) yield the differential forms (10.2.5). Once a measure of phase difference is defined, this form then yields the generalized Cauchy-Riemann equations. The off-diagonal elements of (10.4.5) yield additional constraints which must be satisfied by the singular subspaces, as may be shown by introducing coordinates on $\mathbb{C}P^{n-1}$. We shall not pursue the implications of these constraints at present; in Chapter 12, we shall see an example where they appear to play a role.

To summarize, a generally applicable procedure exists for deriving partial differential equations which must be satisfied by any set of (differentiable) coordinates used to describe a matrix transfer function. The procedure shows precisely how complex analyticity constrains the $2n^2$ real coordinates needed to describe a matrix in $\mathbb{C}^{n \times n}$ to yield only n^2 degrees of freedom.

10.5. Summary

Our first result in this chapter was to derive a set of differential equations (Theorem 10.2.1) which must be satisfied by each pair of singular vectors of a matrix transfer function. These equations are determined, in part, by the motion of the corresponding singular subspaces. There is, however, an additional term which is determined by the differential of the associated singular value (Corollary 10.2.2). Various considerations suggest that the term (10.2.4) is related to some measure of phase difference between the singular vector pair. The fact that $\log \sigma$ cannot be harmonic (Theorem 10.2.3) shows that (10.2.4) cannot be the differential of phase difference.

To investigate these matters further, we returned to the geometric framework we developed in Chapter 9. Using this framework to distinguish the direction of a singular vector from its phase, we were able to derive a set of differential equations resembling the scalar Cauchy-Riemann equations, which relate each singular value to the phase difference between the associated pair of singular vectors (Corollary 10.3.6). The difference between the multivariable equations and the scalar equations is that terms appear in the former due to the motion of the singular subspaces. We show that the effect of the singular subspaces is quantified by certain *connection* forms, which have an interpretation in terms of the fiber bundle structure described in Chapter 9. We also presented a general method (Section 10.4) with which to derive differential equations governing a matrix transfer function in any set of coordinates.

In the next chapter we shall show how our multivariable analogues of the Cauchy-Riemann equations can be turned into multivariable analogues of the gain-phase integral relations.

CHAPTER 11

A MULTIVARIABLE GAIN-PHASE RELATION

11.1. Introduction

In the last three chapters we have seen that each singular value of a matrix transfer function contains additional information about properties of the matrix. Indeed, by exploring the structure of complex projective space, we have shown that each singular value is related to the phase difference between the associated pair of singular vectors. The precise relation is given by a set of "generalized" Cauchy-Riemann equations, which contain additional terms due to motion of the singular subspaces. Unfortunately, it is very difficult to obtain insight using these partial differential equations. It also appears difficult to obtain insight from (10.4.2), the corresponding analogue to the Cauchy Integral Theorem. Hence, our goal in this chapter is to transform the generalized Cauchy-Riemann equations obtained in the previous chapter into a generalization of the scalar gain-phase relation. This generalized relation, to be derived in Section 11.2, shows exactly what additional property of a matrix transfer function is specified once a singular value is known. The integral relation resembles the one relating scalar gain and phase, with an additional term present due to the fact that the singular subspaces are not constant functions of the complex frequency variable. In Section 11.3, coordinates are introduced onto complex projective space so that insight may be gained into how the singular subspaces must move in order that the scalar gain-phase relations be violated. These coordinates are also needed to relate the singular subspaces to the elements of the matrix when expressed in standard coordinates. Section 11.4 contains concluding remarks.

11.2. An Integral Relation Constraining Phase Difference, Singular Values, and Singular Subspaces

From Corollary 10.3.6, the phase difference between a pair of singular vectors is related to the associated singular value through the differential form

$$\frac{\partial[\delta\theta_w(u,v)]}{\partial x}dx + \frac{\partial[\delta\theta_w(u,v)]}{\partial y}dy = \frac{-\partial\log\sigma}{\partial y}dx + \frac{\partial\log\sigma}{\partial x}dy$$

$$-\frac{1}{j}\left[\rho_w^H\frac{\partial[\rho_w(V)]}{\partial x} - \rho_w^H\frac{\partial[\rho_w(U)]}{\partial x}\right]dx \qquad (11.2.1)$$

$$-\frac{1}{j}\left[\rho_w^H\frac{\partial[\rho_w(V)]}{\partial y} - \rho_w^H\frac{\partial[\rho_w(U)]}{\partial y}\right]dy \quad .$$

In principle, this form could be used to compute the value of phase difference at a frequency of interest. All we need do is integrate (11.2.1) along the $j\omega$-axis, yielding

$$\delta\theta_w(u,v)(j\omega_0) - \delta\theta_w(u,v)(-j\infty) = \int_{-\infty}^{\omega_0}\frac{\partial\log\sigma}{\partial x}dy$$

$$-\frac{1}{j}\int_{-\infty}^{\omega_0}\left[\rho_w^H\frac{\partial[\rho_w(V)]}{\partial y} - \rho_w^H\frac{\partial[\rho_w(U)]}{\partial y}\right]dy \qquad (11.2.2)$$

The integral (11.2.2) shows that the phase difference between a pair of singular vectors is determined completely by behavior of the associated singular value and subspaces. This integral is not particularly useful in applications, however, since it does not easily yield insight.

In the scalar case, Cauchy's Integral Theorem may be applied to obtain a more useful statement of the integral relation. Since, as we saw in Section 10.5, the function $\sigma\exp[j\delta\theta_w(u,v)]$ is not generally complex analytic, we must use other mathematical tools. It will turn out that Stokes' Theorem [Spi65] can be used to obtain a useful multivariable "gain-phase" relation. Before proceeding to this result, we need to dispose of some preliminary details.

Recall that a function $f = \mathbb{C} \to \mathbb{C}$ is said to be *conjugate symmetric* if $f(s) = \overline{f(\bar{s})}$. This property is possessed by, for example, rational functions with real coefficients. In our generalization of the scalar result, we shall *assume* that the function $\sigma \exp[j\delta\theta_w(u,v)]$ is conjugate symmetric. This assumption entails no loss of generality for the applications we have in mind. Any local trivialization T_w defined using a real vector w obviously yields a conjugate symmetric measure of phase difference. Furthermore, it is easy to verify that the inner product $u^H v$ is conjugate symmetric; hence the function $\sigma \exp[j\Delta\theta(u,v)]$ defined by (9.3.10)-(9.3.11) also has this property. Note that if $\sigma(0) \neq 0$, then conjugate symmetry implies that $\Delta\theta(u,v)(0) = k\pi$, where k is an odd integer if $\sigma \exp[j\Delta\theta(u,v)] < 0$ and is an even integer otherwise.

The following theorem, although undoubtedly not the most general result possible, will suffice to explain the example in Chapter 12. We shall make some additional assumptions in order to simplify the presentation. First, we shall assume that the matrix transfer function has elements which are rational with real coefficients. (This assumption saves us from having to worry about the effects of time delays, among other things.) Second, we shall assume that the matrix is square and has no poles or transmission zeros in the closed right half plane. The assumption of a square matrix is not restrictive, since we wish to apply this result to the open loop transfer function of a feedback system. As in the scalar case, right half plane poles and zeros may be factored out with a matrix Blaschke product[WDH80].

A more restrictive assumption is that the matrix have distinct singular values at all points of the closed right half plane. We shall also assume that for each singular value there exists a local trivialization whose domain of definition includes both singular vectors associated with that singular value. It should be possible to relax this condition by appropriate use of the δ-transition functions introduced in Section 9.3.

The following theorem may be proven using the generalized Cauchy-Riemann equations (10.3.19) and Stokes' Theorem [Spi65].

Theorem 11.2.1: Let $M(s)$, taking values in $\mathbb{C}^{n \times n}$, have entries which are proper rational functions with real coefficients. Suppose, in addition, that $M(s)$ has the following properties at each point of the closed right half plane:

(i) $M(s)$ is analytic

(ii) $\det M(s) \neq 0$

(iii) $M(s)$ has n distinct singular values.

Let σ be a singular value of M with left and right singular vectors v and u. Assume that

(iv) there exists a local trivialization T_w whose domain of definition includes both v and u at all points of the closed right half plane.

(v) the function $\sigma \exp[j\delta\theta_w(u,v)](s)$ defined using T_w is conjugate symmetric.

In addition, we suppose that

(vi) $\lim\limits_{R \to \infty} \sup\limits_{|\alpha| \leq \pi/2} \left| \log \{\sigma \exp[j\delta\theta_w(u,v)](Re^{j\alpha})\} \right| \cdot 1/R = 0.$

Under these conditions, the phase difference satisfies the following integral relation at each $\omega_0 > 0$

$$\delta\theta_w(u,v)(j\omega_0) - \delta\theta_w(u,v)(0) =$$

$$\frac{1}{\pi}\int_{-\infty}^{\infty}\frac{d\log\sigma}{d\nu}\left[\log\coth\frac{|\nu|}{2}\right]d\nu \tag{11.2.3}$$

$$+\frac{1}{j\pi}\int_{-\frac{\pi}{2}}^{\frac{\pi}{2}}\int_{0}^{\infty}\left\{\rho_w^H\frac{\partial[\rho_w(V)]}{\partial r} - \rho_w^H\frac{\partial[\rho_w(U)]}{\partial r}\right\}dr\,d\alpha$$

where

$$\nu = \log(\omega/\omega_0)$$

and the polar coordinates r and α are defined by

$$s = j\omega_0 + re^{j\alpha}\ ,\quad r \in [0,\infty)\ ,\quad \alpha \in [-\frac{\pi}{2},\frac{\pi}{2}]\ .$$

■

Proof: See Appendix E . ■

Assumptions (iii), (iv), and (vi) appear unnecessarily restrictive; however, they are loose enough to allow analysis of some interesting examples. Our experience has been that the multivariable gain-phase relation holds even if (iii) is violated. Assumption (vi) is a technical assumption needed at one point in the proof. It requires that the function $\sigma\exp[j\,\delta\theta_w(u,v)]$ be sufficiently well behaved near infinity. Scalar rational functions always satisfy this condition and, based upon our experience with examples, it appears reasonable to conjecture that singular values and vectors of matrices of rational functions do also. Rather than seek a more comprehensive theory immediately, we feel that a more profitable goal is to study the results obtained so far to determine whether they are sufficiently useful to justify additional work.

Theorem 11.2.1 is very appealing in that it answers, to the extent possible, the question "What additional property of a matrix transfer function is determined once a singular value gain along the $j\omega$-axis is known?" Equation (11.2.3) shows that each singular value determines the discrepancy between the phase difference between the associated pair of singular vectors and an integral whose value depends upon the behavior of the singular subspaces. The remainder of this section will be devoted to discussing this generalized gain-phase relation in some detail.

Perhaps the most striking thing about the integral relation (11.2.3) is that it decomposes naturally into a term analogous to the scalar case plus an additional multivariable term. Indeed, the first term on the right of this equation is identical to that appearing in the classical relation, with the role of gain played by the singular value. This term shows that a 20N db/decade rate of decrease in the singular value gain contributes 90N° lag to the phase difference $\delta\theta_w(u,v)$. Unlike the scalar case, however, the phase difference is not completely determined by the rate of change in gain. There is an additional contribution due to the way in which directionality properties change with frequency. The latter contribution is quantified by the surface integral in (11.2.3). The value of this integral is completely determined by the motion of the left and right singular subspaces.

Although the term due to directionality properties in (11.2.3) has no direct scalar analogue, it does have an interesting interpretation in terms of the potential theory analogy discussed in Section 10.2. In that section, we showed that $\log\sigma$ does not, in general, satisfy Laplace's equation and therefore cannot be a harmonic function. We gave an interpretation of this fact by showing that the gradient, $\nabla\log\sigma$, behaved like the velocity vector field of a fluid flow in the presence of a *continuous* distribution of sources and sinks. By way of contrast, consider that a scalar rational function violates Bode's relation only if it possesses poles and/or zeros in the right half plane. These poles and zeros, in the potential theory analogy, correspond to discrete sources and sinks, respectively. The effect of these sources and sinks is to produce phase lead or lag in addition to that produced by the rate of gain decrease. The additional term in (11.2.3) may be interpreted as producing additional phase lead or lag even without the presence of right half plane poles or zeros. Instead, the lead or lag is produced by the way in which the singular subspaces, and thus the directionality properties of the system, change with frequency. In Chapter 12, we shall see an example in which this phenomenon appears to cause phase lead to be transferred among the loops of a multivariable system.

The polar coordinate system centered at $j\omega_0$ is very useful, in that the integrand of the surface integral takes a particularly simple form.[32] It is possible to relate this integrand to the Cauchy-Riemann equations written in polar coordinates. If $f(s) = g(s) + jh(s)$ is an analytic function of the complex frequency variable $s = j\omega_0 + re^{j\alpha}$, then it is easy to show that g and h must satisfy

$$\frac{\partial g}{\partial r} = \frac{1}{r}\frac{\partial h}{\partial \alpha} \tag{11.2.4a}$$

$$\frac{1}{r}\frac{\partial g}{\partial \alpha} = -\frac{\partial h}{\partial r} \quad . \tag{11.2.4b}$$

It is also straightforward to show that the integrand of the surface integral satisfies

$$\frac{-1}{j}\left[\rho_w^H \frac{\partial[\rho_w(\mathbf{V})]}{\partial r} - \rho_w^H \frac{\partial[\rho_w(\mathbf{U})]}{\partial r}\right] = \frac{1}{r}\frac{\partial \log\sigma}{\partial \alpha} + \frac{\partial[\delta\theta_w(u,v)]}{\partial r} \quad . \tag{11.2.5}$$

From (11.2.5) and (11.2.4b), it follows that the integrand of the surface integral measures the failure of $\log\sigma + j\delta\theta_w(u,v)$ to satisfy one of the Cauchy-Riemann equations in polar coordinates. In Section 11.3 we shall investigate the way in which the singular subspaces must move

[32] As an examination of the proof of Theorem 11.2.1 will show, the reason for this is the conjugate symmetry inherent in the problem.

to cause this Cauchy-Riemann equation to fail.

Recall that the scalar gain-phase relation contained a weighting function showing that the dependence of phase upon rate of gain change decreases rapidly away from the frequency of interest. A rule of thumb is that the dependence becomes negligible at frequencies more than a decade away.[33] The same weighting function appears in the first integral in (11.2.3), thus showing that the dependence of phase difference at ω_0 upon the rate of change of the singular value at ω decreases as the logarithmic distance between ω and ω_0 increases.

A weighting function is implicit in the second integral as well, although its practical implications are not yet well understood. The reason that a weighting function is implicit in the integral is that the element of surface area in polar coordinates is $r dr d\alpha$. Hence, in effect, the contribution to the value of phase difference at frequency ω_0 due to motion of the singular subspaces in a small piece of the right half plane is inversely proportional to the distance from that piece to the point $s = j\omega_0$. The existence of the weighting function is displayed explicitly when the integral is expressed in standard, rather than polar, coordinates:

$$\begin{aligned} &\frac{1}{\pi}\int_{-\frac{\pi}{2}}^{\frac{\pi}{2}}\int_{0}^{\infty}\frac{1}{r}\left[\rho_w^H\frac{\partial[\rho_w(V)]}{\partial r}-\rho_w^H\frac{\partial[\rho_w(U)]}{\partial r}\right] r dr d\alpha \\ &= \frac{1}{\pi}\int_{-\infty}^{\infty}\int_{0}^{\infty}\frac{1}{r}\left[\rho_w^H\frac{\partial[\rho_w(V)]}{\partial r}-\rho_w^H\frac{\partial[\rho_w(U)]}{\partial r}\right] dx dy \end{aligned} \tag{11.2.6}$$

where

$$r = \sqrt{x^2 + (y - \omega_0)^2} \ .$$

---

[33]There do exist some subtleties related to this point. As discussed in[ZaD63], p.432, there are cases of interest for which the value of phase at a frequency of interest is determined largely by an abrupt change in gain many decades away.

Since the Bode gain-phase relation and the Poisson integrals contain weighting functions, it is perhaps intuitive that the surface integral in (11.2.6) does also. The design implications of the latter weighting function are not yet as well understood as those in the other integrals. However, its existence suggests that an examination of singular subspace behavior in the vicinity of the frequency ω_0 might suffice to determine approximately the effect that the surface integral has upon phase difference at this frequency. This idea will be used heuristically in Chapter 12 to analyze the behavior of an example. In the next section, we shall discuss ways of gaining insight into the integrand of the surface integral, and thus into how the singular subspaces affect the gain-phase relations.

11.3. Coordinates for Complex Projective Space

The purpose of this section is to introduce a set of coordinates onto complex projective space. The one-to-one correspondence which exists between points of $\mathbb{C}P^{n-1}$ and the one-dimensional subspaces of $\mathbb{C}^n$ implies that these coordinates may be used to describe the behavior of the singular subspaces. Such coordinates will prove useful for at least two reasons. First, they will allow us to relate the singular vectors and subspaces to the standard components of a matrix transfer function. Second, the coordinates will allow us to describe the *motion* of the singular subspaces and thus to obtain insight into how that motion causes the scalar gain-phase relation to fail. As the following discussion is somewhat abstract, the reader may wish to only skim this section on first reading, and study the example in Chapter 12 for further motivation.

To simplify the discussion, we shall assume that the local trivialization used to measure phase difference is defined using a *constant* unit vector, which we denote by w_1. Assume that the vectors $w_2, \ldots, w_n$ are chosen so that $\{w_i; i=1, \ldots, n\}$ is a constant orthonormal basis for $\mathbb{C}^n$. This basis will be used to show that the relation between $\log\sigma$ and $\delta\theta_{w_1}(u,v)$ can be influenced by motion of the singular vectors in directions *orthogonal* to that of w_1, the vector used to measure phase difference.

It will frequently be useful to consider a basis $\{w_i\}$ which is a reordering of the standard basis. In particular, trivializations defined using the standard basis vectors are useful in relating the behavior of the singular subspaces to the standard components of the matrix. In many cases phase difference defined using a frequency-dependent local trivialization is also of interest (for example, the measure of phase introduced in Chapter 6 and equations (9.3.10-11)). In such cases δ-transition functions (for example, those defined in Corollary 9.3.2) may be used to relate the value of phase difference obtained using the frequency-dependent trivialization to that obtained using the constant trivialization.

To derive the sets of coordinates in which we are interested, we must first express the left and right singular vectors using the basis $\{w_i\}$:

$$\begin{aligned} v &= \sum_{i=1}^{n} w_i (w_i^H v) \\ u &= \sum_{i=1}^{n} w_i (w_i^H u) \end{aligned} \tag{11.3.1}$$

Using the local trivialization T_{w_1} to define the phase of each vector yields $\theta_{w_1}(v) = \measuredangle w_1^H v$ and $\theta_{w_1}(u) = \measuredangle w_1^H u$. Recall that the number $\theta_{w_1}(\,)$ may be used as *one* of the $2n-1$ real coordinates needed to describe a unit vector in $\mathbb{C}^n$. Since this coordinate distinguishes among all unit vectors spanning the same one-dimensional subspace, the remaining $2n-2$ coordinates are therefore needed only to describe the direction in which the vector lies. We shall now introduce a useful way to define these coordinates.

Denote the magnitude of the ith component of each singular vector by[34]

$$r_i(\mathrm{V}) \triangleq |w_i^H v| \qquad\qquad (11.3.2)$$
$$r_i(\mathrm{U}) \triangleq |w_i^H u|$$

Since $\sum_{i=1}^{n} r_i^2(\mathrm{V}) = \sum_{i=1}^{n} r_i^2(\mathrm{U}) = 1$, it follows that one of the coordinates $\{r_i; i=1, \ldots, n\}$ is redundant. Hence this set of n numbers only conveys $n-1$ pieces of information, and we still need $n-1$ additional coordinates to describe a point of $\mathbb{CP}^{n-1}$. These may be obtained by defining the phase difference between the ith and the 1st components of each vector:[35]

$$\alpha_i(\mathrm{V}) \triangleq \sphericalangle w_i^H v - \theta_{w_1}(v) \qquad\qquad (11.3.3)$$
$$\alpha_i(\mathrm{U}) \triangleq \sphericalangle w_i^H u - \theta_{w_1}(u)$$

Note that $\alpha_1(\mathrm{V}) = \alpha_1(\mathrm{U}) = 0$, by definition. Hence the $n-1$ coordinates $\alpha_i(\cdot)$, $i = 2, \ldots, n$, together with $n-1$ of the coordinates $r_i(\cdot)$, suffice to uniquely determine a point of $\mathbb{CP}^{n-1}$ and thus a one-dimensional subspace of $\mathbb{C}^n$.

We shall now study how the values of the connection forms $\rho_{w_1}^H d\rho_{w_1}(\mathrm{V})$ and $\rho_{w_1}^H d\rho_{w_1}(\mathrm{U})$ are determined by motion of the corresponding singular subspaces. To do this, it

---

[34]When $\{w_i\}$ is the standard basis, the coordinates r_i may be re-written in terms of the latitude angles ϕ_i introduced in Section 9.4. For example, $r_1 = \cos\phi_1$, $r_2 = \sin\phi_1 \cos\phi_2$.

[35]When $\{w_i\}$ is the standard basis, the angles α_i are related to the longitude angles introduced in Section 9.4 via the formula $\alpha_i = \psi_{i-1}$; $i = 2, \ldots, n$.

is convenient to express (11.3.1) using the coordinates (11.3.2-3):

$$v = \exp[j\theta_{w_1}(v)] \cdot \sum_{i=1}^{n} w_i \cdot r_i(\mathrm{V})\exp[j\alpha_i(\mathrm{V})]$$
$$u = \exp[j\theta_{w_1}(u)] \cdot \sum_{i=1}^{n} w_i \cdot r_i(\mathrm{U})\exp[j\alpha_i(\mathrm{U})] \quad . \tag{11.3.4}$$

From the discussion following (9.2.6), it follows easily that the phase-zero reference vectors in each singular subspace are given by

$$\rho_{w_1}(\mathrm{V}) = \sum_{i=1}^{n} w_i \cdot r_i(\mathrm{V})\exp[j\alpha_i(\mathrm{V})]$$
$$\rho_{w_1}(\mathrm{U}) = \sum_{i=1}^{n} w_i \cdot r_i(\mathrm{U})\exp[j\alpha_i(\mathrm{U})] \quad . \tag{11.3.5}$$

These equations, when applied to Corollary 10.3.6, yield

Lemma 11.3.1 With the definitions (11.3.1)-(11.3.3), the differential form (10.3.18) and equations (10.3.19) must satisfy

$$*d\log\sigma = d(\delta\theta_{w_1}\circ(u,v)) + \sum_{i=2}^{n}\{r_i^2(\mathrm{V})d[\alpha_i\circ\mathrm{V}] - r_i^2(\mathrm{U})d[\alpha_i\circ\mathrm{U}]\} \tag{11.3.6}$$

and

$$\frac{\partial\log\sigma}{\partial x} = \frac{\partial[\delta\theta_{w_1}(u,v)]}{\partial y} + \sum_{i=2}^{n}\left[r_i^2(\mathrm{V})\frac{\partial\alpha_i(\mathrm{V})}{\partial y} - r_i^2(\mathrm{U})\frac{\partial\alpha_i(\mathrm{U})}{\partial y}\right]$$
$$\frac{\partial\log\sigma}{\partial y} = -\frac{\partial[\delta\theta_{w_1}(u,v)]}{\partial x} - \sum_{i=2}^{n}\left[r_i^2(\mathrm{V})\frac{\partial\alpha_i(\mathrm{V})}{\partial x} - r_i^2(\mathrm{U})\frac{\partial\alpha_i(\mathrm{U})}{\partial x}\right] \tag{11.3.7}$$

■

Proof: From (11.3.5)

$$d[\rho_{w_i}\circ\mathrm{V}] = \sum_{i=1}^{n} w_i d[r_i\circ\mathrm{V}]\exp[j\alpha_i(\mathrm{V})] + j\sum_{i=1}^{n} w_i r_i(\mathrm{V})\exp[j\alpha_i(\mathrm{V})]d[\alpha_i\circ\mathrm{V}] \quad .$$

Hence

$$\rho_{w_i}^H(V)d[\rho_{w_i}\circ V] = \sum_{i=1}^{n} r_i(V)d[r_i\circ V] + j\sum_{i=1}^{n} r_i^2(V)d(\alpha_i\circ V)$$
$$= j\sum_{i=2}^{n} r_i^2(V)d(\alpha_i\circ V) \quad .$$

The last equality follows since $\alpha_1(V) \equiv 0$, by definition, and since

$$2\sum_{i=1}^{n} r_i(V)d(r_i\circ V) = d[\sum_{i=1}^{n} r_i^2(V)] = d[1] = 0 \quad .$$

Equation (11.3.6) follows from this result and an analogous one for the right singular subspace. Equations (11.3.7) follow by an argument identical to that in the proof of Corollary 10.3.6.

■

Equations (11.3.7) show how the failure of the Cauchy-Riemann equations is related to the coordinates $r_i(\cdot)$ and $\alpha_i(\cdot)$. Suppose that we introduce polar coordinates onto the complex plane, as we did in deriving (11.2.7-8). Then we can also show how the failure of the scalar gain-phase relation is related to the behavior of $r_i(\cdot)$ and $\alpha_i(\cdot)$. Indeed, the integrand of the surface integral in (11.2.3) becomes

$$\rho_{w_1}^H \frac{\partial[\rho_{w_1}(V)]}{\partial r} - \rho_{w_1}^H \frac{\partial[\rho_{w_1}(U)]}{\partial r} = j\sum_{i=2}^{n}\left[r_i^2(V)\frac{\partial\alpha_i(V)}{\partial r} - r_i^2(U)\frac{\partial\alpha_i(U)}{\partial r}\right] . \tag{11.3.8}$$

As we discussed in the previous section, this surface integral quantifies the extent to which motion of the singular subspaces affects the value of phase difference between the singular vectors.

As we saw in the previous section, the surface integral in (11.2.3) quantifies the extent to which motion of the singular subspaces affects the value of the phase difference between the singular vectors. Unfortunately, no generally useful method for approximating the value of this integral is currently available. However, recalling our earlier conjecture, it appears possible that useful qualitative information may be obtained by examining the integrand (11.3.8) in the vicinity of a frequency at which we wish to approximate the value of phase.

To simplify the notation, we shall assume that the basis $\{w_i\}$ is equal to the standard basis $\{e_i\}$, with the usual ordering. Although our results can readily be extended to other bases, use of the standard basis has an advantage in that we can think in terms of the physical loops of the system. Indeed, let σ denote a singular value of a stable matrix transfer function evaluated at $s = j\omega_0$, and let u and v denote the associated pair of right and left singular vectors. Suppose that u has been chosen so that $u = \rho_{w_1}(\mathrm{U})$ (and, therefore, $v = \rho_{w_1}(\mathrm{V})\exp[j\delta\theta_{w_1}(u,v)]$). Then an input

$$u(t) = ue^{j\omega_0 t} = \begin{bmatrix} r_1(\mathrm{U}) \\ r_2(\mathrm{U})\exp[j\alpha_2(\mathrm{U})] \\ \vdots \\ r_n(\mathrm{U})\exp[j\alpha_n(\mathrm{U})] \end{bmatrix} e^{j\omega_0 t} \tag{11.3.9a}$$

produces a steady-state output

$$v(t) = \sigma v \exp[j(\omega_0 t + \delta\theta_1(u,v))] = \begin{bmatrix} r_1(\mathrm{V}) \\ r_2(\mathrm{V})\exp[j\alpha_2(\mathrm{V})] \\ \vdots \\ r_n(\mathrm{V})\exp[j\alpha_n(\mathrm{V})] \end{bmatrix} \sigma\exp[j(\omega_0 t + \delta\theta_1(u,v))] \tag{11.3.9b}$$

From these expressions for $u(t)$ and $v(t)$, it is easy to see that the coordinates $r_i(\mathrm{V})$ and $r_i(\mathrm{U})$ determine how the magnitude of each signal is distributed among the loops. The coordinates $\alpha_i(\mathrm{V})$ and $\alpha_i(\mathrm{U})$ determine the relative phase difference between those components of the signals appearing in the first loop and the components appearing in other loops. As we shall see in the next chapter, it is sometimes possible to determine how the singular subspaces are related to the standard components of the underlying matrix. By exploiting this relation to estimate the values of $r_i(\cdot)$ and $\alpha_i(\cdot)$, it will then be possible to estimate the effect that loop coupling has upon the multivariable gain-phase relation. We shall now discuss two conditions which must be satisfied if the singular subspaces are to affect the multivariable gain-phase relation by causing nonzero values of the integrand (11.3.8).

First, note that the integrand (11.3.8) will be nonzero at a frequency only if at least one of the singular vectors has a component orthogonal to e_1, the first standard basis vector. Indeed, if both singular vectors are aligned with e_1 at all frequencies, then the function $\sigma \exp[j\delta\theta_{e_1}(u,v)]$ is everywhere equal to the (1,1) element of the matrix transfer function. Since this element is analytic, it follows that the scalar gain-phase relation is satisfied in this case. Even if the singular vectors are only approximately aligned with e_1, then it is reasonable that the scalar relation should be satisfied approximately. Alternately, we could say that the scalar gain-phase relation can be violated only if the behavior of the singular value and vectors depends *significantly* upon more than one element of the transfer function matrix.

A second condition necessary for the integrand (11.3.8) to be nonzero at a frequency is that at least one of the angles $\alpha_i(V)$ and/or $\alpha_i(U)$ have a nonzero partial derivative at that frequency. Equivalently, this says that the phase difference between the 1st components of a pair of singular vectors can deviate from the value it would take in the scalar case *only* if the relative phase between the 1st and the ith component of one of the vectors changes with frequency. The presence of the coefficient $r_i^2(\cdot)$ implies that the size of the component whose relative phase changes must be sufficiently large for the effect to be appreciable. Additional insight into this behavior, as well as a tentative interpretation, may be gleaned from the example to be discussed in Chapter 12. For the case $n=2$, some visual insight may be gained by identifying the coordinates $r_i(\cdot)$ and $\alpha_i(\cdot)$ with the latitude and longitude angles depicted in

Figure 9.4.1.

In previous chapters we have seen that the measure of phase difference which appears to have the greatest physical significance is that obtained by using the right singular vector to define a local trivialization (equations (9.3.10-11)). The preceding discussion will apply to this case directly when the right singular subspaces are approximately constant; simply let these constant values be used to define basis vectors $\{w_i\}$. Otherwise, as we pointed out at the beginning of this section, δ-transition functions may be used to relate this measure of phase difference to one defined using a constant trivialization.

Finally, as we remarked at the close of the previous section, the fact that the generalized gain phase relation (11.2.3) depends upon the value of (11.3.8) at all points of the right half plane greatly complicates the analysis of the integral. Nevertheless, the weighting function implicit in the integral suggests that an analysis of the integrand (and hence of the coordinates $r_i(\cdot)$ and $\alpha_i(\cdot)$) in the vicinity of a frequency of interest may suffice to allow useful information to be obtained. In the next chapter, we shall analyze an example for which this conjecture appears to be true.

11.4. Summary and Conclusions

Our goal in this and the three preceding chapters has been to derive a multivariable version of the classical gain-phase relation. The role of gain in our generalization is played by a singular value, while that of phase is played by the phase difference between the associated pair of singular vectors. Motivated by the scalar case, we conjectured that the function $\sigma\exp[j\delta\theta_w(u,v)]$ might be complex analytic. However, as we have now seen, this function violates all the standard scalar results, including the Cauchy-Riemann equations and Cauchy's Integral Theorem as well as the Bode gain-phase relation. In each case, failure of the classical result is due to the motion of the singular subspaces and hence to the way in which directionality properties of the system change with frequency. Our results show that the discrepancy between $\sigma\exp[j\delta\theta_w(u,v)]$ and an analytic function may be quantified using the connection forms $\rho_w^H d\rho_w(V)$ and $\rho_w^H d\rho_w(U)$.

The multivariable gain-phase relation is important for several reasons. In principle, it shows that knowledge of each singular value gain *does* contain additional information about properties of a matrix transfer function. Hence imposing a design specification upon a singular value of a transfer function *implicitly* constrains other properties of the underlying system. Since such constraints may conflict with other design specifications, it is important to determine precisely what they are. Toward that end, it is interesting that the additional property constrained by a singular value is the phase difference between its associated pair of singular vectors. It is important to note that the constraint imposed also depends upon the behavior of the singular subspaces, and hence upon inherently directional properties of the system.

Clearly, much work remains before the information provided by the multivariable gain-phase relation can be fully incorporated into analysis and design. For instance, the unnecessarily restrictive assumptions in Theorem 11.2.1 need to be removed. However, the most critical need is to provide system-theoretic interpretations of the integral and of the quantities which it constrains. It is also necessary to have ways of relating the singular values and vectors to the physical components of the system. In the next chapter, we shall present a simple, academic, example where the generalized gain-phase relation appears to quantify a potentially *beneficial* effect of coupling in a multivariable system.

CHAPTER 12

AN EXAMPLE ILLUSTRATING THE MULTIVARIABLE GAIN-PHASE RELATION

12.1. Introduction

The primary purpose of this chapter is to construct an example tying together several of the results of this monograph, including the multivariable gain-phase relation developed in Chapter 11. Because the phenomena which we wish to discuss are complex, and not yet well understood, we shall make every effort to study the simplest possible nontrivial case. For this reason, our example will be the transfer function of a system with only two inputs and two outputs. As a secondary goal, we also wish to demonstrate that multivariable systems can possess certain properties which differ significantly from those of scalar systems. We shall now describe our motivation for the type of example we consider, and outline briefly the steps we take in its construction.

Recall the Bode gain-phase relation which, as discussed in Chapter 2, implies the existence of a tradeoff between sensitivity reduction at low frequencies and robustness against high frequency modelling error. An alternate statement is that there exists a tradeoff between the phase margin of a feedback system and the rate at which gain can be rolled off near crossover. Hence, the fact that gain and phase of a scalar transfer function cannot be manipulated independently manifests itself as a tradeoff between feedback system properties in different frequency ranges. In particular, there may exist a large peak in the sensitivity function at gain crossover. As the references cited in Chapter 5 show, essentially the same tradeoff is also present in multivariable systems, *provided* that the gain in all loops is constrained to cross over at the same frequency and rate. These straightforward generalizations of classical results to multivariable systems are most useful when directionality and structure do not play a role in design. Frequently, however, multivariable design specifications do exhibit structure, perhaps due to some feedback sensors becoming noisy at lower frequencies than others, or due to the presence of structured modelling error (Section 5.4). As we saw in Section 7.6, these design specifications can be loosely translated into requirements that the open loop transfer function have small gain in the direction of the structured noise or uncertainty; gain in the other directions can remain large to achieve other design goals. With this observation, it appears reasonable to ask how fast one loop of a multivariable system can be rolled off, provided that gain in the other loops is allowed to remain large and to cross over at a higher frequency. Although this question is strictly academic, similar questions do arise from structured design specifications; in particular, it is of interest to determine whether loop coupling can play a beneficial or a harmful role in meeting such specifications. A complete answer to these problems is as yet unavailable; however, our results should prove useful in their eventual resolution.

In this chapter we will construct a matrix transfer function with the property that gain in one loop rolls off at 40 *db/decade* near crossover, but whose sensitivity function has a peak at crossover substantially smaller than that which would be present in the scalar case. Our strategy will be to require gain in the other loop to remain large, and to introduce an appropriate type of beneficial coupling between the high and low gain loops. We shall also investigate some harmful effects that coupling has at other frequencies, and postulate a tradeoff between these two effects.

The analysis in this chapter will also serve to illustrate the utility of several of our previous results. Of particular use are the approximations of feedback properties from those of the open loop system (Chapter 7) and the multivariable gain-phase relation (Chapter 11). We will also develop procedures for approximating the singular values and vectors from the standard components of the open loop transfer function. By using *only* the assumption that there exists a spread in open loop gains, it will be possible to approximate several properties of the singular value decomposition.

Our example will illustrate that the effects of loop coupling may be used to perform design tradeoffs in ways having no direct scalar analogue. Since the example does not arise from a physical system, however, we cannot offer any physical interpretations of the phenomena we observe. Nevertheless, the fact that loop coupling can have such an effect is quite interesting, and may have potential applications in the future. In any event, our primary purposes in this chapter are to illustrate the multivariable gain-phase relation and to develop some methods for analyzing the effects of coupling in an existing system. For the present, we are only secondarily interested in speculations about the practical applications of coupling, and our example should be considered in this context.

In discussing our example, we shall need to develop several results which are *generally* applicable to systems with two inputs and outputs. Since these results are of independent interest, we have organized the chapter by separating these general remarks from those pertaining specifically to the example. In Section 12.2 we review the parameterization of systems with two inputs and outputs which we originally introduced in Section 9.5. We then show that five of these eight parameters can be easily approximated whenever a spread in loop gains exists and is sufficiently large. Section 12.3 is devoted to preliminary construction of our example. We review the results from Chapter 7 concerning feedback systems with a large spread in loop gain and find that feedback properties depend critically upon a phase parameter which cannot easily be approximated using the methods of Section 12.2. Section 12.4 is devoted to a specialization of the multivariable gain-phase relation (11.2.6) to systems with two inputs and outputs. We show how insight may be obtained into the effects that loop coupling has upon the relation between a singular value gain and the phase difference between the associated pair of singular vectors. In Section 12.5 we complete the construction of our example, using the results obtained in Section 12.4 to choose loop-coupling terms which affect feedback properties at crossover in a desirable way. Finally, Section 12.6 contains a summary and directions for further research.

12.2: Approximation of the Singular Value Decomposition for a System with a Spread in Loop Gain

Let $L(s)$ be a transfer function taking values in $\mathbb{C}^{2\times 2}$ and denote the elements of L and its singular value decomposition by

$$\begin{aligned} L &= \begin{bmatrix} L_1 | L_2 \end{bmatrix} \\ &= \begin{bmatrix} L_{11} & L_{12} \\ L_{21} & L_{22} \end{bmatrix} \\ &= V \Sigma U^H \\ &= \begin{bmatrix} v_1 & v_2 \end{bmatrix} \begin{bmatrix} \sigma_1 & 0 \\ 0 & \sigma_2 \end{bmatrix} \begin{bmatrix} u_1^H \\ u_2^H \end{bmatrix} \\ &= \begin{bmatrix} v_{11} & v_{12} \\ v_{21} & v_{22} \end{bmatrix} \begin{bmatrix} \sigma_1 & 0 \\ 0 & \sigma_2 \end{bmatrix} \begin{bmatrix} u_{11} & u_{12} \\ u_{21} & u_{22} \end{bmatrix}^H \end{aligned} \tag{12.2.1}$$

In Section 9.5 we introduced a parameterization of the singular value decomposition of L. For convenience, we shall restate the result as a lemma.

Lemma 12.2.1 Let s be a frequency for which $L(s)$ in (12.2.1) has

(i) no poles

(ii) no transmission zeros

(iii) distinct singular values ($\sigma_1 > \sigma_2$)

(iv) singular vectors which satisfy $v_{11} \neq 0$ and $u_{11} \neq 0$.

Then the transfer function L evaluated at frequency s may be parameterized as

$$L = \begin{bmatrix} \cos\phi_v & -\sin\phi_v e^{-j\psi_v} \\ \sin\phi_v e^{j\psi_v} & \cos\phi_v \end{bmatrix} \cdot \begin{bmatrix} \sigma_1 e^{j\delta\theta_1(1)} & 0 \\ 0 & \sigma_2 e^{j\delta\theta_2(2)} \end{bmatrix} \cdot \begin{bmatrix} \cos\phi_u & -\sin\phi_u e^{-j\psi_u} \\ \sin\phi_u e^{j\psi_u} & \cos\phi_u \end{bmatrix}^H , \tag{12.2.2}$$

where $\phi_v, \phi_u \in [0, \frac{\pi}{2})$ and $\delta\theta_1(1), \delta\theta_2(2), \psi_v, \psi_u \in [-\pi, \pi)$.

■

We shall often denote the largest and smallest singular values by $\overline{\sigma} = \sigma_1$ and $\underline{\sigma} = \sigma_2$, respectively, and use a corresponding notation for the associated pairs of singular vectors.

Suppose that gain in the first loop of this system is much larger than that in the second loop; i.e., suppose that[36]

$$||L_1|| \gg ||L_2|| \quad . \tag{12.2.3}$$

We shall show that this assumption *alone* allows us to approximate *five* of the eight parameters in (12.2.2). Note that some of the following discussion parallels that of Section 7.5.

It follows from the min-max property of singular values that

$$\overline{\sigma}[L] \geq ||L_i|| \geq \underline{\sigma}[L] \quad . \tag{12.2.4}$$

Hence, at frequencies for which (12.2.3) holds, assumption (iii) of Lemma 12.2.1 is satisfied.

We shall next use (12.2.3)-(12.2.4) to show that the right singular subspaces of L are approximately aligned with the standard basis directions. A basic property of singular values

---

[36] As elsewhere, $||\cdot||$ denotes either the standard Euclidean vector norm or its induced matrix norm.

is that

$$Lu_i = \sigma_i v_i \tag{12.2.5}$$

and from this fact it follows that

$$\sigma_1 v_1 = u_{11} L_1 + u_{21} L_2 \tag{12.2.6a}$$

$$\sigma_2 v_2 = u_{12} L_1 + u_{22} L_2 \tag{12.2.6b}$$

Hence

$$\begin{aligned} \sigma_1 &\le |u_{11}| \cdot \|L_1\| + |u_{21}| \cdot \|L_2\| \\ &\le |u_{11}| \cdot \|L_1\| + \|L_2\| \quad . \end{aligned} \tag{12.2.7}$$

Together (12.2.4) and (12.2.7) imply

$$1 - \frac{\|L_2\|}{\|L_1\|} \le |u_{11}| \quad . \tag{12.2.8}$$

Thus, whenever (12.2.3) holds, it follows from (12.2.8) that $|u_{11}| \approx |u_{22}| \approx 1$, $|u_{12}| \approx |u_{21}| \ll 1$, and

$$\phi_u \ll 1 \quad , \tag{12.2.9}$$

so that the right singular subspaces are indeed aligned with the standard basis directions.

The spread in loop gain (12.2.3) also allows the left singular vector and the largest singular value to be easily approximated. Using (12.2.3) and (12.2.9) in (12.2.6a) yields

$$Lu_1 = \sigma_1 v_1$$

$$= \sigma_1 \exp[j\,\delta\theta_1(1)] \begin{bmatrix} \cos\phi_v \\ \sin\phi_v e^{j\psi_v} \end{bmatrix}$$

$$\approx L_1$$

From this approximation it follows readily that

$$\sigma_1 \approx ||L_1|| \qquad (12.2.10)$$

$$\delta\theta_1(1) \approx \measuredangle L_{11} \qquad (12.2.11)$$

$$\phi_v \approx arctan \frac{|L_{21}|}{|L_{11}|} \qquad (12.2.12)$$

$$\psi_v \approx \measuredangle \frac{L_{21}}{L_{11}} \quad . \qquad (12.2.13)$$

Based solely upon assumption (12.2.3), we have shown that it is possible to approximate five of the eight parameters in (12.2.2). Approximation of the remaining three is more problematic without additional knowledge of the functions L_{ij} and will be deferred until later.

12.3: Preliminary Construction of Example

Consider a scalar feedback system whose open loop transfer function has the form $g(s) = \frac{c}{(s+.01)^2}$. At break frequency, this function has 90° phase lag; as $\omega \to \infty$, the phase lag asymptotically approaches 180°. If the gain c is too large, then the phase lag at crossover will cause a large peak in the sensitivity function. To illustrate, if

$$g(s) = \frac{012}{(s+.01)^2} \quad , \tag{12.3.1}$$

then crossover occurs near $\omega = .1\ rad/sec$. The phase at crossover is approximately $-170°$, and there is a $15db$ peak in the sensitivity function.

We will now construct a matrix transfer function whose smallest singular value is approximately equal to $g(s)$. We will also require the largest singular value to roll off at a slower rate and cross over at a significantly higher frequency. This will allow us to investigate whether coupling between high and low gain loops can be used to significantly reduce the peak in the sensitivity function from the value it attains in the scalar case.[37]

We shall first require that the gain in the second loop of the system be equal to that of (12.3.1); this can be done by choosing

$$L_{12}(s) = \frac{c_{12}}{(s + .01)^2} \tag{12.3.2}$$

$$L_{22}(s) = \frac{c_{22}}{(s + .01)^2} \quad , \tag{12.3.3}$$

where $\sqrt{c_{12}^2 + c_{22}^2} = .012$. Since all we know about σ_2 is that $\sigma_2 \leq ||L_2||$, there is no guarantee at this point that $\sigma_2 = |g|$, as we desire. This condition will be enforced by appropriately specifying the constants c_{i2} later in this chapter.

Our next step is to insure that the first loop of the system rolls off more slowly, and at a higher frequency, so that a large spread in loop gain will be present at frequencies near crossover for the second loop.

---

[37]Of course, for a complete analysis we would also need to determine the effect of coupling upon other system properties, such as those quantified by the complementary sensitivity function.

It is easy to verify that choosing

$$L_{11}(s) = \frac{10}{s + .01} \tag{12.3.4}$$

implies that $||L_1(s)|| > ||L_2(s)||$ throughout the right half plane. Furthermore, the spread in loop gain increases away from the origin, being bounded below by

$$\begin{aligned} ||L_1(s)|| \Big/ ||L_2(s)|| &\geq |L_{11}(s)| \Big/ ||L_2(s)|| \\ &= 833 \cdot |s + .01| \quad . \end{aligned} \tag{12.3.5}$$

The fact that the spread in loop gain becomes large at frequencies below crossover for the second loop allows us to use (12.2.9) and (12.2.11) to conclude that

$$\phi_u \ll 1 \tag{12.3.6}$$

and

$$\delta\theta_1(1) \approx \sphericalangle \frac{10}{j\omega + .01} \tag{12.3.7}$$

for those frequencies.

Before proceeding with our example, we need to recall the relation between open and closed loop properties of a system with a spread in loop gain. First, recall definitions (6.5.1-2)

$$\phi(i) = arccos\ |u_i^H v_i|$$

$$\Delta\theta(i) = \sphericalangle u_i^H v_i$$

and observe that for systems with two inputs and outputs the angles $\phi(1)$ and $\phi(2)$ are equal. Hence, we define

$$\phi \triangleq \phi(1) \equiv \phi(2) \quad . \tag{12.3.8}$$

We showed in Chapter 7 that, at frequencies for which $\sigma_1 \gg 1/\cos\phi$, the system sensitivity function is approximately equal to

$$S_{app} = e^{j\Delta\theta(2)} u_2 (\cos\phi + \sigma_2 e^{j\Delta\theta(2)})^{-1} v_2^H \quad . \tag{12.3.9}$$

At frequencies for which $\sigma_1 \gg 1/\cos\phi$ and $\cos\phi \gg \sigma_2$, the sensitivity and complementary sensitivity functions are approximately equal to

$$S_{app} = e^{j\Delta\theta(2)} u_2 (\frac{1}{\cos\phi}) v_2^H \tag{12.3.10a}$$

$$T_{app} = e^{-j\Delta\theta(1)} v_1 (\frac{1}{\cos\phi}) u_1^H \quad . \tag{12.3.10b}$$

Finally, at frequencies for which $\cos\phi \gg \sigma_2$, the complementary sensitivity function is approximately equal to

$$T_{app} = \sigma_1 v_1 (1 + \cos\phi \sigma_1 e^{j\Delta\theta(1)})^{-1} u_1^H \quad . \tag{12.3.11}$$

We can use (12.3.9-11) to guide us in selecting the function $L_{ij}(s)$. Clearly, we must first determine the values of σ_2, $\Delta\theta(i)$, and the coupling parameter ϕ. However, as we discussed at the close of Section 9.4, the fact that $\phi_v \gg 1$ implies that $\Delta\theta(i) \approx \delta\theta_i(i)$ and it is also easy to show that $\phi \approx \phi_v$. Hence there is no loss of information incurred by using the parameter $\phi_v \approx \arctan |L_{21}/L_{11}|$ to measure the strength of coupling between the high and low gain directions and the parameters $\delta\theta_i(i)$ to measure the phase difference between each pair of singular vectors.

We already have an approximation for $\delta\theta_1(1)$, and can determine ϕ_v by fixing the gain of L_{21}. Approximating the value of the phase parameter $\delta\theta_2(2)$ proves to be more difficult; however, at least in principle, $\delta\theta_2(2)$ can be determined from the multivariable gain-phase relation (11.2.6). To obtain additional insight, we shall now specialize this relation to systems with two inputs and outputs.

12.4: The Multivariable Gain-Phase Relation for a System with Two Inputs and Outputs

Consider a system with two inputs and outputs parameterized as in (12.2.2). Using this parameterization and the multivariable gain-phase relation (11.2.6), we shall now show that the function $L_{21}(s)$ can have a significant effect upon the phase parameter $\delta\theta_2(2)$.

Suppose that assumptions (i)-(iv) of Lemma 12.2.1 are satisfied throughout the closed right half plane. Then assumptions (i)-(iii) of Theorem 11.2.2 are satisfied and we can use the local trivialization $T_w = T_{e_i}$ to measure phase difference between the ith pair of singular vectors. Hence assumptions (iv) and (v) of Theorem 11.2.1 also hold. The fact that the left (right) singular vectors form an orthonormal basis yields the parameterizations (12.2.2) of L and (9.5.7-8) of the reference vectors $\rho_{e_i}(V_i)$ and $\rho_{e_i}(U_i)$. The latter parameterizations, when substituted into the multivariable gain-phase relation (11.2.6), yield the following corollary.

Corollary 12.4.1: Let $L(s)$, taking values in $\mathbb{C}^{2\times 2}$, have entries which are proper rational functions with real coefficients. Suppose, in addition, that $L(s)$ has the following properties at each point of the closed right half plane:

(i) $L(s)$ is analytic

(ii) $\det L(s) \neq 0$

(iii) $L(s)$ has distinct singular values.

(iv) $|u_{11}| \neq 0$ and $|v_{11}| \neq 0$

In addition, we suppose that, for $i=1,2$,

(v) $$\lim_{R \to \infty} \sup_{|\alpha| \le \frac{\pi}{2}} \left| \log \{\sigma_i \exp[j\delta\theta_i(i)](Re^{j\alpha})\} \right| \cdot 1/R = 0$$

Under these conditions, the phase difference between each pair of singular vectors must satisfy the following integral relation at each $\omega_0 > 0$

$$\delta\theta_i(i)(j\omega_0) - \delta\theta_i(i)(0) =$$

$$\frac{1}{\pi}\int_{-\infty}^{\infty}\frac{d\log\sigma_i}{d\nu}\left[\log\coth\frac{|\nu|}{2}\right]d\nu \tag{12.4.1}$$

$$-\frac{(-1)^i}{\pi}\int_{-\frac{\pi}{2}}^{\frac{\pi}{2}}\int_0^{\infty}[\sin^2\phi_v\frac{\partial\psi_v}{\partial r} - \sin^2\phi_u\frac{\partial\psi_u}{\partial r}]dr\, d\alpha$$

where $\nu = \log(\omega/\omega_0)$ and the polar coordinates r and α are defined from $s = j\omega_0 + re^{j\alpha}$, $r \in [0,\infty]$, $\alpha \in [-\frac{\pi}{2},\frac{\pi}{2}]$.

■

One useful and interesting observation can be obtained immediately from (12.4.1). Since the sets $\{v_i\}$ and $\{u_i\}$ must each form an orthonormal basis for $\mathbb{C}^2$, it follows that one of the left (respectively, right) singular subspaces can change with frequency only if the other does also. Hence the functions $\sigma_i\exp[j\delta\theta_i(i)]$ are, in effect, coupled. They violate the scalar gain-phase relation by amounts which are equal in magnitude but opposite in sign. A very tentative interpretation of this result is that, due to the effects of loop coupling, phase lag or lead is transferred from one of these functions to the other.[38]

It is clear from the approximation (12.3.9) that the phase parameter $\Delta\theta(2)$ (which approximately equals $\delta\theta_2(2)$) is important in determining whether good feedback properties are present near crossover frequency for the smallest singular value. Hence we would like to use (12.4.1) to investigate the effects of loop coupling upon the value of $\delta\theta_2(2)$. Before we can

---

[38]See also Section 11.3.

apply this integral relation, however, we must determine whether assumptions (i)-(v) are satisfied. Since this necessarily depends upon the specific example considered, we defer this discussion until the next section.

If the assumptions in Corollary 12.4.1 are satisfied, then the value of $\delta\theta_2(2)$ can be approximated provided that we can approximate the two integrals in (12.4.1). The first integral is similar to the familiar Bode integral, and its value can be approximated once the value of $\frac{d\log\sigma_2}{d\nu}$ is known. We will return to this in the next section.

We shall now study the surface integral in (12.4.1). Using the assumption that a spread in loop gain exists at frequencies of interest, we shall show that qualitative information about the effect of this integral can be obtained from the functions $L_{11}(s)$ and $L_{21}(s)$. Since the integral quantifies the extent to which the functions $\sigma_i\exp[j\delta\theta_i(i)]$ are coupled and therefore violate the scalar gain-phase relation, any information we can obtain about its behavior will prove useful when completing our example in the next section.

Although the integral of interest has domain of integration equal to the right half plane, not all points of the half plane contribute equally to its value. As we saw in Section 11.2 (see the discussion surrounding (11.2.9)), the element of surface area in the plane is $rdrd\alpha$, and this fact implies that the integrand of the surface integral contains a weighting function equal to 1/r. Indeed, this weighting function shows that the contribution to the value of $\delta\theta_i(i)(j\omega_0)$ from the integrand

$$C(s) \triangleq \sin^2\phi_v \frac{\partial\phi_v}{\partial r} - \sin^2\phi_u \frac{\partial\phi_u}{\partial r} \quad , \tag{12.4.2}$$

when integrated over a small piece of the right half plane, is inversely proportional to the distance of that piece to the frequency ω_0. Motivated by this observation, we propose to study the integral of $C(s)$ over the right half plane by looking only at the integral of $C(s)$ over the small semicircular region

$$D_\varepsilon(\omega_0) \triangleq \{s = j\omega_0 + re^{j\alpha} : \alpha \in [-\frac{\pi}{2}, \frac{\pi}{2}],\ r \in [0,\varepsilon]\} \tag{12.4.3}$$

which is weighted most heavily in the integral. Equivalently, we shall attempt to gain insight into the value of

$$I(j\omega_0) \triangleq \int_0^\infty \int_{-\frac{\pi}{2}}^{\frac{\pi}{2}} (\sin^2\phi_v \frac{\partial \psi_v}{\partial r} - \sin^2\phi_u \frac{\partial \psi_u}{\partial r})d\alpha dr \tag{12.4.4}$$

by looking instead at the integral

$$I_\varepsilon(j\omega_0) \triangleq \int_0^\varepsilon \int_{-\frac{\pi}{2}}^{\frac{\pi}{2}} (\sin^2\phi_v \frac{\partial \psi_v}{\partial r} - \sin^2\phi_u \frac{\partial \psi_u}{\partial r})d\alpha dr \quad . \tag{12.4.5}$$

Our primary concerns are in approximations and qualitative insights; we wish to know when $I(j\omega_0)$ is apt to have large magnitude and whether the sign of $I(j\omega_0)$ is likely to be positive or negative. The latter information is particularly important since it determines which of the functions $\sigma_i \exp[j\delta\theta_i(i)]$ will experience additional lag and which will experience additional lead.

At present, we cannot provide a firm theoretical basis for analyzing $I(j\omega_0)$ by looking only at the local behavior of the integrand (12.4.2). Nevertheless, this technique seems to work well when applied to examples. It appears reasonable to conjecture that if the integrand (12.4.2) varies sufficiently smoothly, then at least the sign of (12.4.4) will agree with that of (12.4.5).

We now describe a method for approximating the value of the integral (12.4.5). We shall assume that the spread in loop gain (12.2.3) is sufficiently large that approximations (12.2.9) and (12.2.12-13) hold arbitrarily well.

First, the assumption that $\phi_u \ll 1$ suggests that the term $\sin^2\phi_u \frac{\partial \psi_u}{\partial r}$ can be neglected in (12.4.5). If we make the additional assumption that $\phi_v(s)$ is approximately constant over $D_\varepsilon(\omega_0)$, then

$$I_\varepsilon(j\omega_0) \approx \sin^2\phi_v(j\omega_0) \int_0^\varepsilon \int_{-\frac{\pi}{2}}^{\frac{\pi}{2}} \frac{\partial \psi_v}{\partial r}\, d\alpha dr \quad . \tag{12.4.6}$$

Our next step is to use (12.2.12-13) to approximate the interior integral in (12.4.6). Define the ratio

$$R(s) \triangleq L_{21}(s)/L_{11}(s) \quad , \tag{12.4.7}$$

and recall from (12.2.12-13) that

$$\phi_v \approx \arctan |R| \tag{12.4.8}$$

$$\psi_v \approx \measuredangle R \quad . \tag{12.4.9}$$

It follows that we should be able to approximate the integral of $\frac{\partial \psi_v}{\partial r}$ by summing over the contributions to this integral from each of the poles and zeros of $R(s)$. Denote these by $\{p_i;\ i=1, \ldots, N_p\}$ and $\{z_i;\ i=1, \ldots, N_z\}$, respectively. Hence, with $s = j\omega_0 + re^{j\alpha}$,

$$\int_{-\frac{\pi}{2}}^{\frac{\pi}{2}} \frac{\partial \psi_v}{\partial r}\, d\alpha = \sum_{i=1}^{N_z} \int_{-\frac{\pi}{2}}^{\frac{\pi}{2}} \frac{\partial \measuredangle(z_i - s)}{\partial r}\, d\alpha - \sum_{i=1}^{N_p} \int_{-\frac{\pi}{2}}^{\frac{\pi}{2}} \frac{\partial \measuredangle(p_i - s)}{\partial r}\, d\alpha \quad . \tag{12.4.10}$$

Assuming, for simplicity, that the poles and zeros of $R(s)$ are *real*, it follows that

$$\frac{\partial \measuredangle(z_i - s)}{\partial r} = \frac{-z_i \sin\alpha - \omega_0 \cos\alpha}{(z_i - r\cos\alpha)^2 + (\omega_0 + r\sin\alpha)^2} \tag{12.4.11}$$

and

$$\frac{\partial \sphericalangle(p_i - s)}{\partial r} = \frac{-p_i \sin\alpha - \omega_0 \cos\alpha}{(p_i - r\cos\alpha)^2 + (\omega_0 + r\sin\alpha)^2} \quad . \tag{12.4.12}$$

If we assume that $r < \varepsilon \ll \omega_0$, then the terms containing r in (12.4.11-12) may be ignored, and the integral of each term is approximately

$$\int_{-\frac{\pi}{2}}^{\frac{\pi}{2}} \frac{\partial \sphericalangle(z_i - s)}{\partial r}\, d\alpha \approx \frac{-2\omega_0}{z_i^2 + \omega_0^2} \tag{12.4.13}$$

or

$$\int_{-\frac{\pi}{2}}^{\frac{\pi}{2}} \frac{\partial \sphericalangle(p_i - s)}{\partial r}\, d\alpha \approx \frac{-2\omega_0}{p_i^2 + \omega_0^2} \quad . \tag{12.4.14}$$

Finally, combining (12.4.6), (12.4.10), and (12.4.13-14) yields

$$I_\varepsilon(j\omega_0) \approx \varepsilon \cdot \sin^2\phi_v(j\omega_0) \cdot \left\{ \sum_{i=1}^{N_z} \frac{-2\omega_0}{z_i^2 + \omega_0^2} + \sum_{i=1}^{N_p} \frac{2\omega_0}{p_i^2 + \omega_0^2} \right\} , \tag{12.4.15}$$

where, from (12.4.8), it follows that

$$\sin^2\phi_v(j\omega_0) = |R(j\omega_0)|^2 \Big/ (1 + |R(j\omega_0)|^2) \quad . \tag{12.4.16}$$

To summarize, we have developed an approximation to the surface integral in (12.4.1) based upon the behavior of the integrand (12.4.2) near the frequency of interest. Furthermore, we have assumed that the transfer function has a spread in loop gain, as does our example. Hence approximations (12.2.9,12,13) allow us to approximate the effect of coupling using the ratio (12.4.7) and (12.4.15-16). The latter equations show that $I_\varepsilon(j\omega_0)$ will be large *only* if the ratio (12.4.7) is large. Furthermore the sign of the ratio depends upon the relative locations of the poles and zeros of (12.4.7). To illustrate, suppose that $R(s) = k(s+z)/(s+p)$. If $0 \le z < p$, (so that $R(s)$ is a *lead* filter) then $I_\varepsilon(j\omega_0) < 0$. Hence, from (12.4.1), we should expect that the function $\sigma_2 \exp[j\delta\theta_2(2)]$ would experience additional phase *lead*, while the function $\sigma_1 \exp[j\delta\theta_1(1)]$ would experience additional *lag*. If $0 \le p < z$, then opposite conclusions would be drawn.

12.5: Final Construction of Example

We shall now use the results of the previous section to complete the construction of our example. In order to do this, we need to insure that assumptions (i)-(v) of Corollary 12.4.1 are satisfied. As it turns out, the structure of the class of systems we are considering (i.e., plants with a large spread in loop gain) tends to guarantee that most of these assumptions will be satisfied *independently* of our choices of the constants c_{i2} in (12.3.2-3) and the function $L_{21}(s)$. Clearly, (i) will be satisfied if we choose L_{21} to be stable. By (12.2.4) and (12.3.5), assumption (iii) is satisfied *regardless* of our choice of L_{21}. By combining (12.2.8) and (12.3.5), it follows easily that the first part of assumption (iv) is satisfied. It turns out that the second part of (iv) will also be satisfied whenever the ratio $|L_{12}/L_{11}|$ is sufficiently small. Indeed, from (12.2.6a), we have

$$\sigma_1 v_{11} = u_{11} L_{11} + u_{21} L_{12}$$

which implies that

$$\begin{aligned} \sigma_1 |v_{11}| &\ge \Big| |u_{11}| \cdot |L_{11}| - |u_{21}| \cdot |L_{12}| \Big| \\ &= |u_{11}| \cdot |L_{11}| \cdot \Big| 1 - \tan\phi_u \, \Big|\frac{L_{12}}{L_{11}}\Big| \Big| \quad . \end{aligned} \tag{12.5.1}$$

By (12.2.8) and (12.3.5), $\tan\phi_u < \tan 28° < .54$, and by (12.3.5) $|L_{12}/L_{11}| \le ||L_2|| \Big/ |L_{11}| \le 1/8.33$. Using these results in (12.5.1) yields $|v_{11}| > 0$ as desired. Hence each member of the pair of singular vectors (u_i, v_i) lies within the domain of definition of the local trivialization T_{e_i} and assumption (iv) of Corollary 12.4.1 is satisfied. As we discussed in Chapter 11, condition (v) appears to be satisfied by virtue of the fact that $L(s)$ has rational entries. Finally, we need to verify that $L(s)$ is minimum phase, so that (ii) is satisfied. To do this, we shall need additional information about the function $L_{21}(s)$.

To complete our example, we must now choose the constants c_{i2} in (12.3.2-3) and the function $L_{21}(s)$ so that

(a) $L(s)$ is minimum phase (hence (ii) is satisfied),

(b) the smallest singular value rolls off at $\approx 40 db/decade$ near crossover,[39]

(c) near crossover, the surface integral contributes sufficient phase *lead* to $\sigma_2 \exp[j\delta\theta_2(2)]$ that the attendant peak in $\bar{\sigma}[S]$ is reduced from its value in the scalar case.

We shall approach these tasks by first choosing $L_{21}(s)$ so that the surface integral in (12.4.1) contributes phase lead to $\delta\theta_2(2)$ near $\omega = .1 rad/sec$. This procedure will fix the poles and zeros of $L_{21}(s)$, leaving only a constant factor in this term and the constants c_{i2} undetermined. These constants will then be chosen to insure that (a) is satisfied, that σ_2 crosses over near $\omega = .1 rad/sec$, and that a sufficient amount of phase lead is produced so that the peak in sensitivity is reduced.

Our results from the preceding section suggest that we should attempt to satisfy (c) by choosing the ratio $R = L_{21}/L_{11}$ so that the integral (12.4.6) has a relatively large *negative* value at $\omega = .1$. From (12.4.15), we see that if $R(s) = z - s$, then the value of (12.4.6) will indeed be negative. However, if $R(s)$ contains no poles, then the measure of coupling $\phi = \phi_v$ will

[39]Recall that crossover occurs at frequencies for which $\sigma_2 = \cos\phi$.

increase to $\pi/2$ at high frequencies. Since we know from (12.3.10) that this will tend to cause poor feedback properties, additional poles should be introduced into $R(s)$ so that coupling does not become excessively large. Hence we choose

$$R(s) = \frac{ks}{(s+1)^2} \tag{12.5.2}$$

where k is a constant whose value is to be determined. Using this value of $R(s)$ in (12.4.10-14) yields

$$\int_{-\frac{\pi}{2}}^{\frac{\pi}{2}} \frac{\partial \psi_v}{\partial r} \, d\alpha \approx \frac{-2}{\omega_0} + \frac{4\omega_0}{1+\omega_0^2} \Big|_{\omega_0=.1} \tag{12.5.3}$$

$$\approx -19.6 \quad .$$

If our attempt to estimate the sign of the surface integral (12.4.4) by that of (12.4.5) is valid, then from (12.5.3) it follows that choosing $|k|$ sufficiently large in (12.5.2) should cause additional phase lead to be added to $\delta\theta_2(2)$ near $\omega{=}.1 rad/sec$. In view of (12.3.9), this is precisely what is needed to reduce the peak in sensitivity near crossover. On the other hand, $|R(j\omega)|$ has a peak at $\omega = 1 rad/sec$. If this peak is too large then, by (12.3.10), feedback properties will be poor near this frequency. Hence we should expect that our ability to use coupling to produce phase lead near $\omega{=}.1 rad/sec$ may be compromised by the need to limit the effects of coupling at higher frequencies.

Combining (12.3.2-4), (12.4.7), and (12.5.2), our tentative open loop transfer function becomes

$$L(s) = \begin{bmatrix} \dfrac{10}{s+.01} & \dfrac{c_{12}}{(s+.01)^2} \\ \\ \dfrac{10ks}{(s+.01)(s+1)^2} & \dfrac{c_{22}}{(s+.01)^2} \end{bmatrix} \tag{12.5.4}$$

where $c_{12}^2 + c_{22}^2 = (.012)^2$. We still need to choose the constants k and c_{i2} so that $L(s)$ is minimum phase and $\sigma_2 = ||L_2||$. By examining $\det L(s)$, it is easy to see that constraining $c_{12} \geq 0$, $c_{22} > 0$, and $k \leq 0$ will guarantee the minimum phase property.

Note that if $c_{12}=0$, then $\bar{\sigma}[S] \geq 1/|1+L_{22}|$ and scalar analysis methods show that there will exist at least a $15db$ peak in sensitivity near the frequency at which $|L_{22}| = 1$. Hence we must choose $c_{12} \neq 0$; as an initial try, we will set $c_{12} = c_{22} = .012/\sqrt{2} \approx .0085$, and adjust these values later, if necessary.

Our final step is to iterate on the value of k in (12.5.2). This procedure resulted in the following final value for the open loop transfer function:

$$L(s) = \begin{bmatrix} \dfrac{10}{s+.01} & \dfrac{.0085}{(s+.01)^2} \\ \\ \dfrac{-38s}{(s+.01)(s+1)^2} & \dfrac{.0085}{(s+.01)^2} \end{bmatrix} . \tag{12.5.5}$$

In Section 12.2, we developed approximations (12.2.9-13) for five of the eight parameters describing $L(s)$. Using Bode gain and phase plots, it is straightforward to verify that these approximations hold very well. In particular, our choice of the ratio $R(s)$ (12.4.7) has the desired effect upon the parameters ϕ_v and ψ_v (12.4.8-9). Furthermore, as we discussed following (12.3.9-11), the fact that $\phi_u \ll 1$ implies that the parameters used in the equations may be approximated by $\phi \approx \phi_v$ and $\Delta\theta(i) \approx \delta\theta_i(i)$. We shall now discuss the behavior of these parameters in detail.

Figure 12.5.1 contains plots of the singular values of $L(s)$; for purposes of comparison, the magnitudes of the eigenvalues are also shown. Note that the smallest singular value rolls off at $40db/decade$, as desired. Indeed, a comparison of Bode gain plots shows that $\sigma_2[L]$ is very nearly equal to the gain in the second loop, and thus to that of $g(s)$ (12.3.1). Hence the amount of phase contributed to $\delta\theta_2(2)$ by the first integral in (12.4.1) is equal to $\measuredangle g$; near $\omega = .1 rad/sec$, $\measuredangle g \approx -170°$.

The actual values of $\delta\theta_i(i)$ are plotted in Figure 12.5.2, along with $\measuredangle g$ for purposes of comparison. The phases of the eigenvalues of $L(s)$ are also shown, although they are indistinguishable from the phase parameters. We see that $\delta\theta_2(2)$ is less negative than $\measuredangle g$ at frequencies below $\omega = 1 rad/sec$. In particular, near $\omega = .1 rad/sec$, the crossover frequency for the second loop, $\delta\theta_2(2) \approx -150°$; hence the coupling between loops contributes 20° phase *lead* to the value of this parameter. On the other hand, the value of $\delta\theta_1(1)$ is more negative than it would be if $\sigma_1 \exp[j\delta\theta_1(1)]$ were a scalar transfer function; this is because the peak in σ_1 near $\omega = 1 rad/sec$ has no apparent effect upon $\delta\theta_1(1)$.[40] The fact that loop coupling causes one of the functions $\sigma_i \exp[j\delta\theta_i(i)]$ to experience additional lead and the other to experience additional lag is clear from our discussion following Corollary 12.4.1 .

The discrepancies between the phase parameters $\delta\theta_i(i)$ and the values they would take if the $\sigma_i \exp[j\delta\theta_i(i)]$ were analytic functions are caused by the coupling between the directions of the singular subspaces. The strength of this coupling is quantified by the parameter $\phi = arccos\,|u_i^H v_i|$, which is approximately equal to ϕ_v. Both these parameters are plotted in Figure 12.5.3 (their values are indistinguishable), along with the value of $arctan\,|L_{21}/L_{11}|$. Note that the coupling angle becomes large (> 60°) near $\omega = 1 rad/sec$, in between the crossover frequencies for the two singular values.

---

[40]That the gain of L_{21} can affect σ_1, but not $\delta\theta_1(1)$, is obvious from (12.2.10-11).

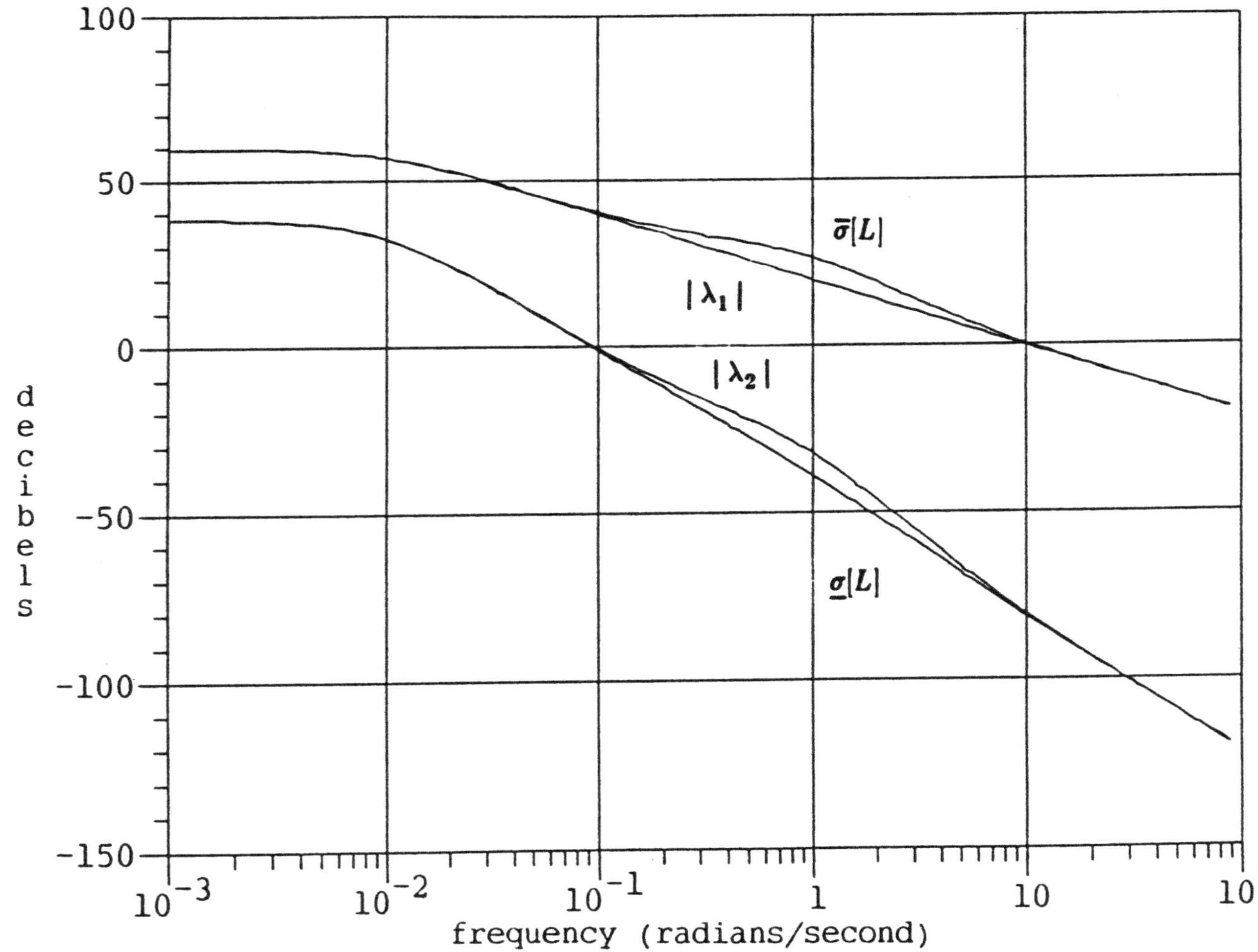

Figure 12.5.1: Open Loop Singular Values and Eigenvalue Magnitudes

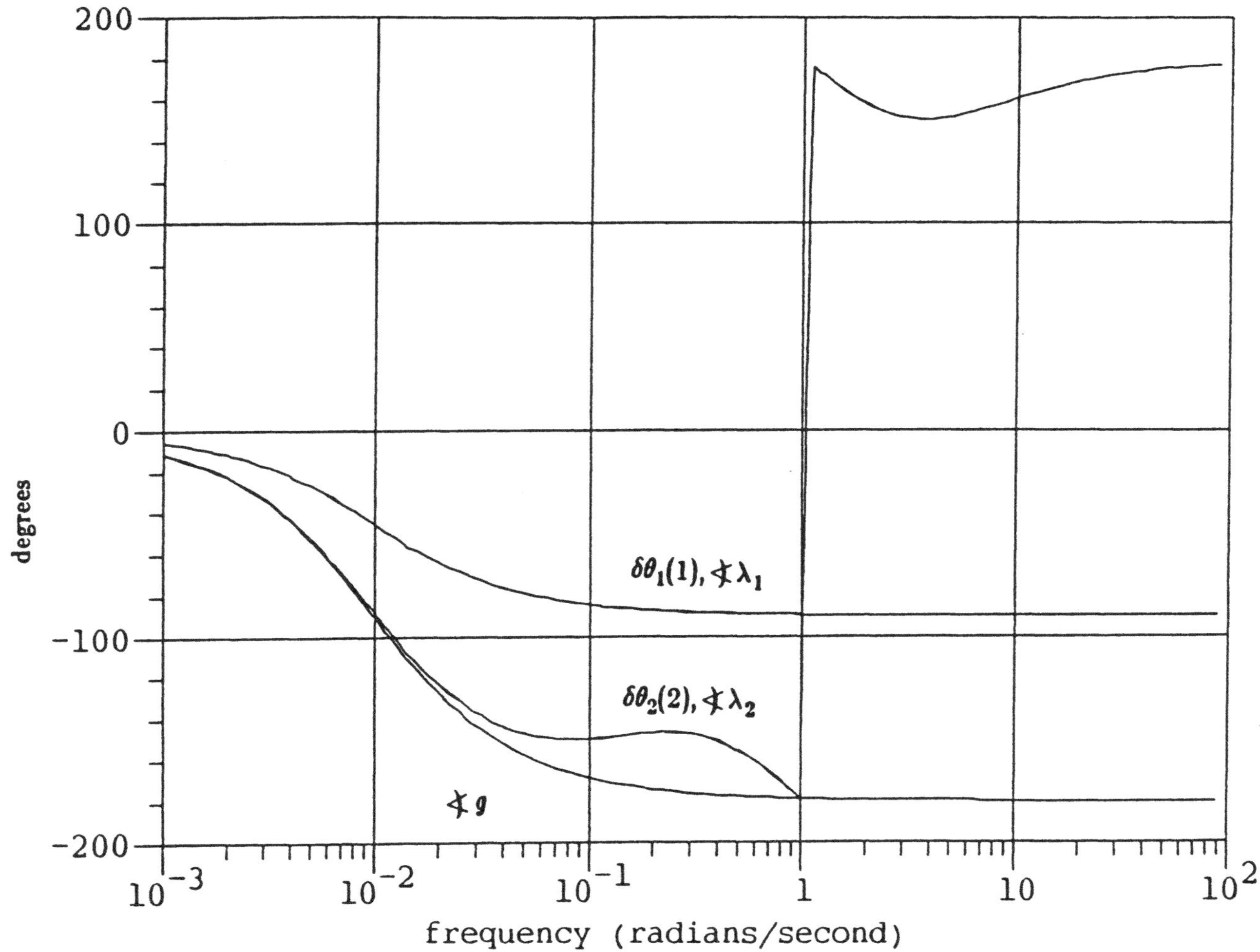

Figure 12.5.2: Phase Parameters

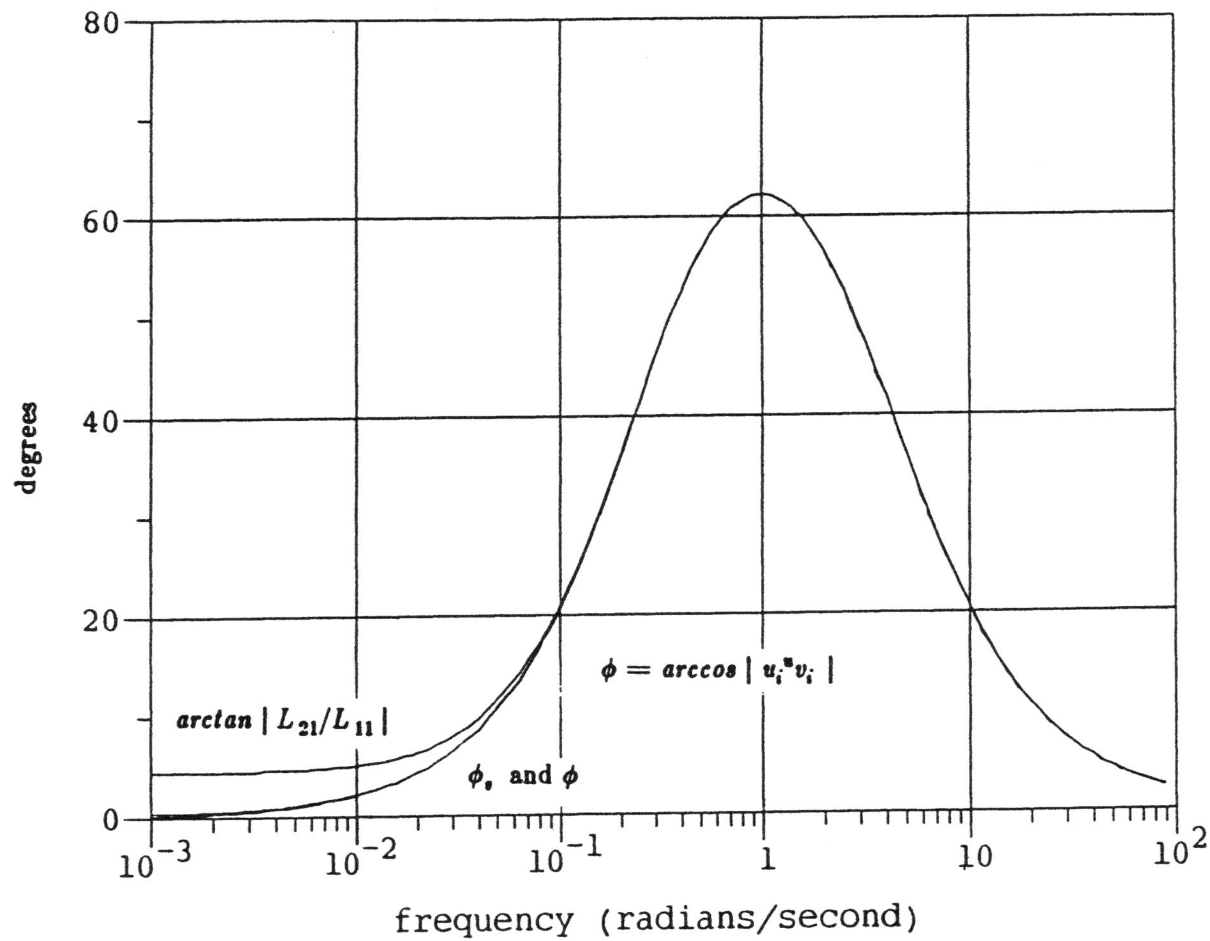

Figure 12.5.3: The Angle Between Singular Subspaces, and Approximations

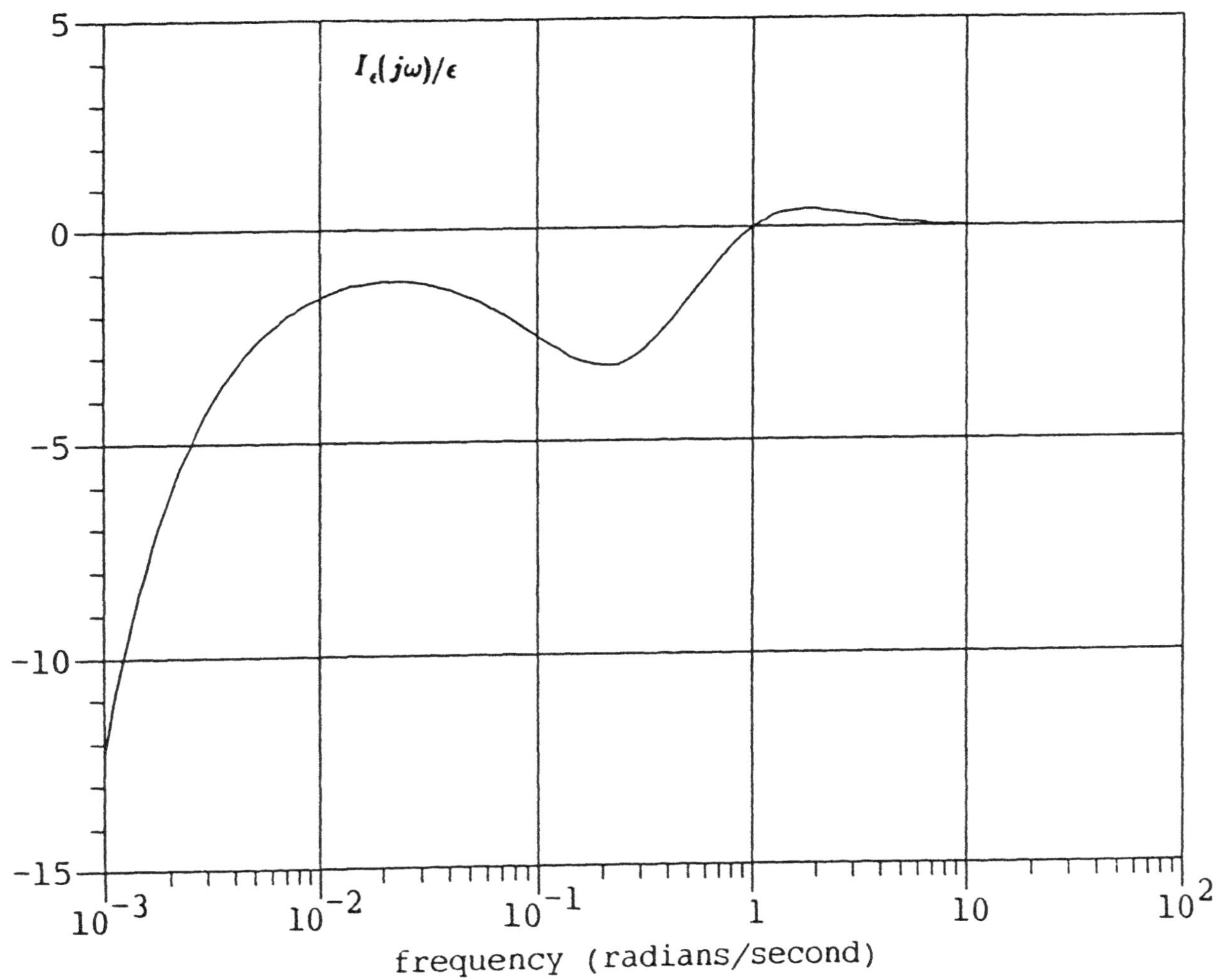

Figure 12.5.4: Approximation (12.5.6) for Surface Integral

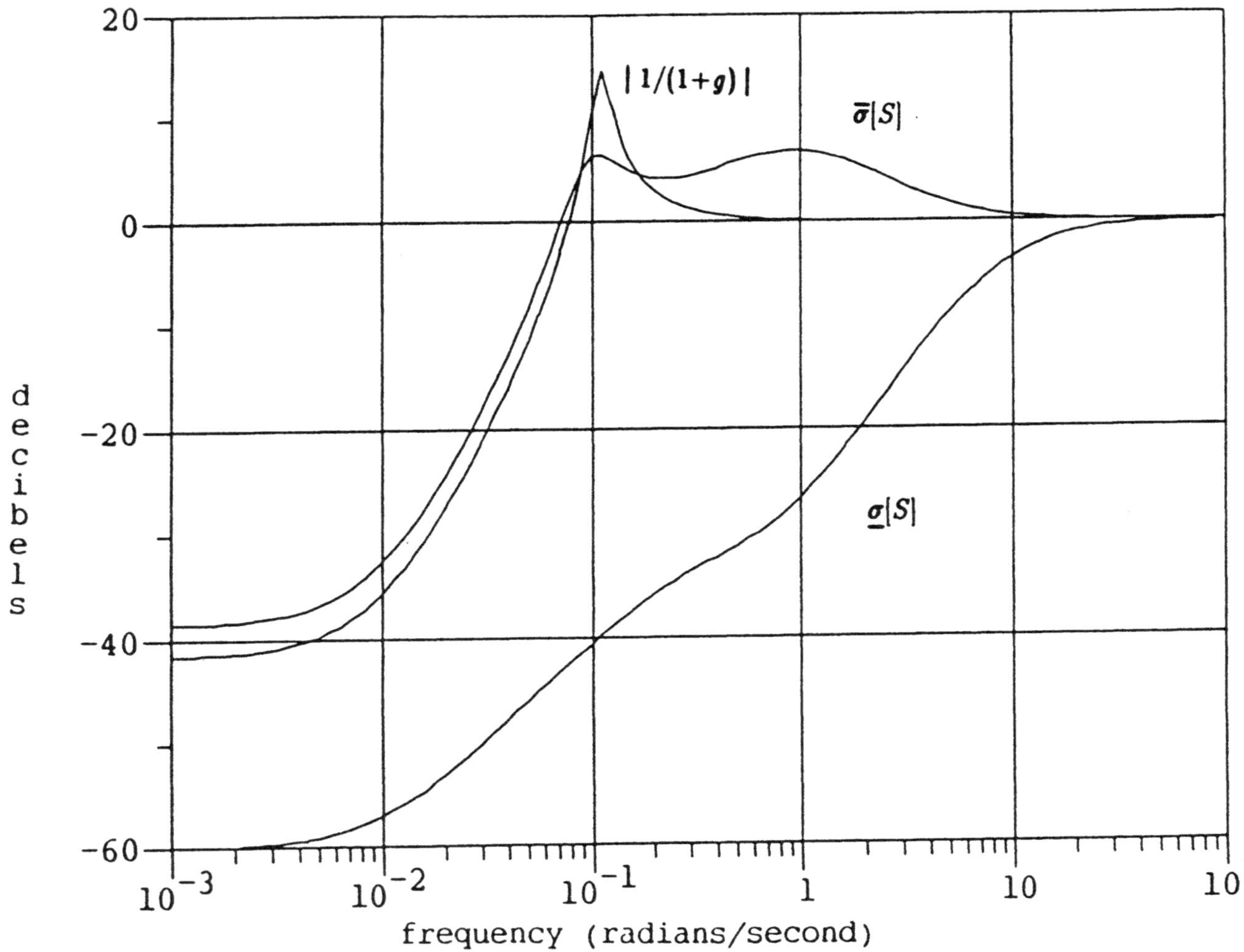

Figure 12.5.5: Singluar Values of Sensitivity Function

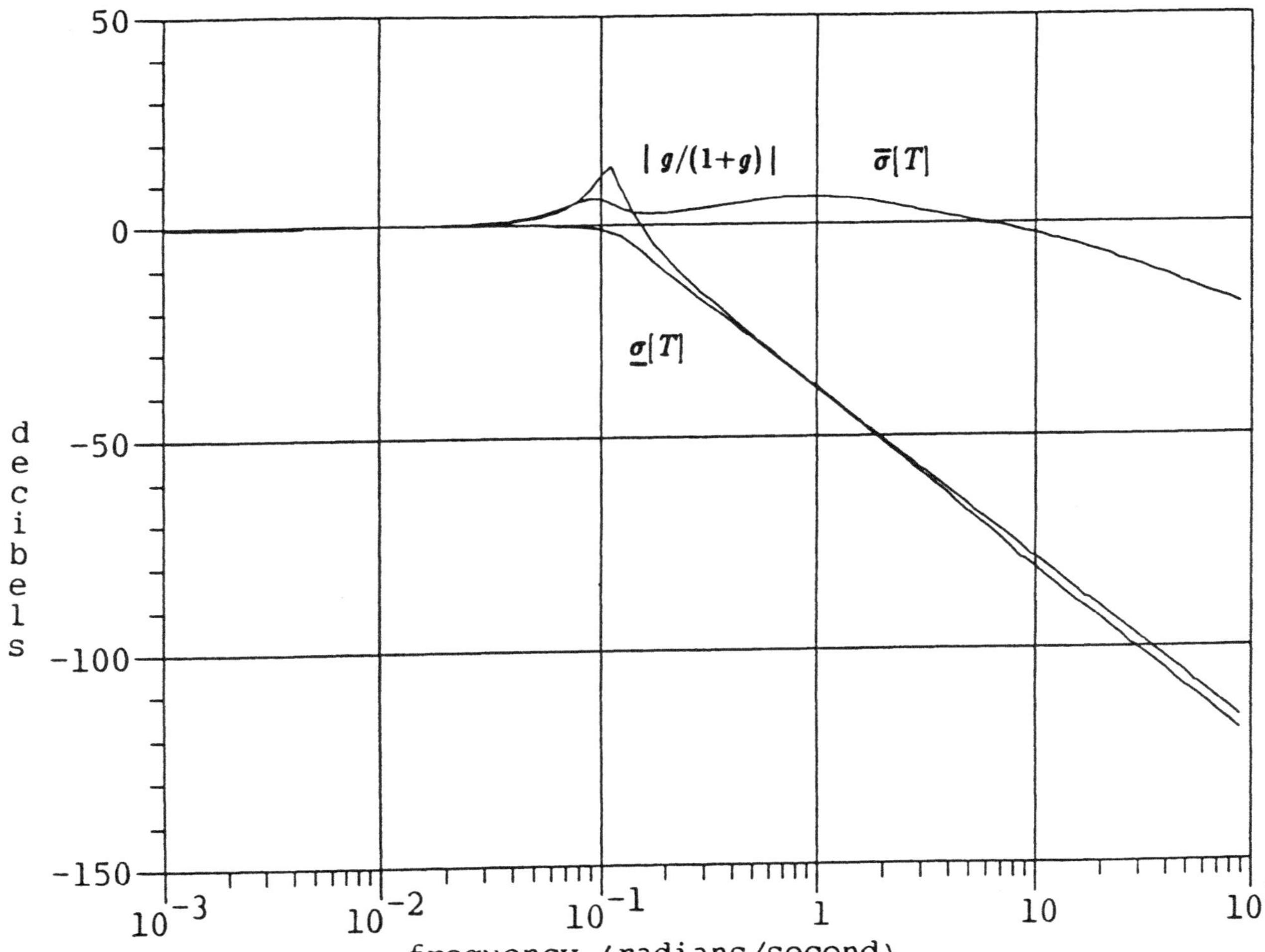

Figure 12.5.6: Singular Values of Complementary Sensitivity Function

The effect that coupling between singular subspaces has upon the value of $\delta\theta_2(2)$ is quantified by the surface integral $I(j\omega_0)$ (12.4.4) and may be estimated from (12.4.15); specifically,

$$I_\varepsilon(j\omega_0)/\varepsilon = \sin^2\phi_v(j\omega_0)\left[-\frac{2}{\omega_0}+\frac{4}{1+\omega_0^2}\right] . \tag{12.5.6}$$

This approximation is plotted in Figure 12.5.4. This plot indicates that $\delta\theta_2(2)$ should be *less* negative than $\measuredangle g$ at frequencies below $\omega = 1 rad/sec$ and *more* negative at higher frequencies. This prediction is confirmed by Figure 12.5.2. Hence the heuristic methods we used to evaluate the effects of coupling, and to choose the poles and zeros of $L_{21}(s)$, work very well in this example.

We now proceed to analyze the sensitivity properties of a feedback system for which $L(s)$ is the open loop transfer function. First, using the eigenvalue (or characteristic loci) stability robustness criterion[PoM79], it is easy to verify that such a feedback system is stable.[41]

Since $\Delta\theta(i) = \delta\theta_i(i)$, and since $\phi = \phi_v$, it follows that Figures 12.5.1-3 can be used together with approximations (12.3.9-11) to estimate sensitivity properties. As we discussed in Chapter 7, this information can be used to predict that $\bar{\sigma}[S]$ should have two peaks. The first of these will occur at crossover frequency for $\sigma_2[L]$; it is easy to verify that $\sigma_2 = \cos\phi$ at a frequency near $\omega = .1 rad/sec$. Since $\sigma_2[L]$ is only slightly less than one at crossover, and since $\Delta\theta_2(2) = -150°$ at this frequency, the peak in $\bar{\sigma}[S]$ should be substantially less than that of a scalar feedback system with open loop transfer function $g(s)$ (12.3.1). On the other hand, there will also be a peak in sensitivity between the crossover frequencies for the two loops,

[41] It may be verified that the eigenvalues of $L(s)$ have no right half plane branch points. Hence they must obey the scalar gain phase relations, as is evidenced by comparing Figures 12.5.1 and 12.5.2.

where coupling becomes large ($\phi > 60°$).[42] Figure 12.5.5 shows that the sensitivity function behaves as we predict. Indeed, the plots of $\bar{\sigma}[S]$ and approximation (12.3.9) are indistinguishable on this scale. There exist two peaks in $\bar{\sigma}[S]$, both approximately equal to $6.5db$. The first peak is near $\omega = .11 rad/sec$; note that it is substantially smaller than the $15db$ peak in the scalar sensitivity function $1/(1+g)$. The second peak is also approximately $6.5db$, and occurs near $\omega = 1 rad/sec$, the frequency at which the coupling angle ϕ is largest.

That the two peaks in $\bar{\sigma}[S]$ have equal magnitude is no accident, as the parameter k in (12.5.4) was selected with this in mind. In fact, there appears to be a tradeoff between the sizes of these two peaks. Increasing $|k|$, and thus the strength of coupling, causes a further decrease in the size of the first peak, at the expense of a larger second peak. Decreasing $|k|$, on the other hand, has exactly the opposite effect. This behavior is consistent with our discussion following (12.5.3). An interesting question is whether the tradeoff observed here is an *inherent* property of linear systems, or whether it is specific to a particular class, which includes this example. For the present example, the tradeoff discussed is clear, since ϕ_v and ψ_v may be identified with the gain and phase of the scalar transfer function $R(s)$. It should be possible to obtain some insight into the general case using the additional equations constraining matrix transfer functions (in particular, their singular subspaces) which we described in Section 10.4.

Finally, the singular values of the complementary sensitivity function are shown in Figure 12.5.6. For purposes of comparison, we also plot the scalar complementary sensitivity function $g/(1+g)$. As one would expect from the identity $S + T = I$, $\bar{\sigma}[T]$ also has two peaks, each approximately 6.5 db, while the scalar function has a 15 db peak near crossover. Note that $\underline{\sigma}[T]$ rolls off at 40 *db/decade* starting near $\omega = .1 rad/sec$, near crossover frequency

[42]As was done in Section 7.5, we can approximate $\bar{\sigma}[S]$ near the crossover frequency for $\sigma_1[L]$ by using

for $\underline{\sigma}[L]$. It may also be verified that the right singular subspaces of T become closely aligned with the standard basis directions as $\underline{\sigma}[T]$ rolls off. Indeed, at $\omega = 1 rad/sec$, the discrepancy between these subspaces is less than one degree.

12.6. Conclusions

It is clear from our example that multivariable feedback systems can possess properties substantially different from those of their scalar counterparts.[43] Furthermore, at least in simple cases, it appears possible to use the results of Chapter 7 and 11 to analyze and to manipulate these properties. In particular, we have seen that loop coupling may be used to obtain additional flexibility in design, by allowing tradeoffs to be performed between feedback properties in different directions as well as in different frequency ranges. Hence, it is tempting to conjecture that incorporating loop coupling into a multivariable system could allow otherwise unattainable design goals to be achieved. Since, at the most, coupling will only allow tradeoffs to be performed in different ways, rather than to disappear completely, a pressing question is whether or not the potential benefits of coupling will be outweighed by the potential drawbacks. Resolving this issue will require us to consider a more complete design specification than that discussed above. For example, greater attention must be paid to the directionality properties of the complementary sensitivity function.

If it turns out that the net effect of strong loop coupling is always detrimental, then design specifications should incorporate a constraint that the system be sufficiently decoupled. On the other hand, if loop coupling proves to have advantages in certain cases, then methods must be found to introduce coupling into a design. We avoided this issue in constructing our

---

(7.4.12).

[43]Another example, whose behavior is even more dramatic, may be found in [FrL83b].

example by designing the open loop transfer function directly. In practice, freedom to completely specify this function might not be available. For example, compensators which completely invert the plant and substitute desired dynamics are known to face potential robustness difficulties. Hence, if we are to vary the amount of coupling present in a system, we may need to introduce compensation internal to the plant. Alternately, requirements for coupling (or decoupling) could be imposed upon the initial design of the plant. The analysis techniques demonstrated in this chapter should prove useful for all these tasks.

CHAPTER 13

CONCLUSIONS

The practice of feedback design is subject to many limitations and tradeoffs. Among the sources of these are constraints imposed by the structure of the feedback loop, by realizability, and by properties of the plant to be controlled. Consequently, there is a need for a well developed theory describing these limitations and an accompanying set of heuristics to aid in analysis and design. This monograph contributes to the body of such a theory in several ways. First, we have presented a view of the classical theory of scalar feedback systems which highlights the role played by constraints upon feedback properties. The latter manifest themselves either as algebraic tradeoffs between system properties at a single frequency, or as analytic tradeoffs between feedback properties in different frequency ranges. We then examined the extension of these ideas to multivariable systems. Our second major result concerned the relation between open and closed loop properties of a multivariable feedback system. This relation is useful in understanding the algebraic design tradeoffs present in multivariable design. A third contribution was the development of a multivariable extension of the scalar gain-phase integral relation. This extension was based upon a careful analysis of how standard results from complex variable theory appear after frequency dependent transformations to coordinates, such as singular values, useful in feedback analysis and design. The final major contribution was an application of all the results of the monograph to explain the behavior of an interesting multivariable transfer function. Our results were used both to construct the example transfer function as well as to predict the resulting feedback properties of the system. This application required the development and use of several heuristics, and served to integrate the results of the monograph.

Clearly, much additional work remains before the design limitations in a multivariable feedback system can be fully understood and incorporated into the design process. For example, as we noted in Chapter 5, the feedback properties of a multivariable system can differ significantly at different loop-breaking points. The results of Chapter 7 can be used to analyze this difficulty at a single frequency. However, such an analysis is yet incomplete, and must be extended to account for tradeoffs between system properties in different frequency ranges. The integral relation developed in Chapter 11 should prove useful in analyzing such tradeoffs. In addition, the mathematical foundation laid in Chapters 8-10 should prove useful in obtaining multivariable generalizations of the other integral relations that were examined in Chapter 3. In particular, a multivariable generalization of the Poisson integral relation would aid in understanding the design limitations imposed by directionally dependent bandwidth constraints and by multivariable transmission zeros.

We expect that the analysis tools developed in this monograph can be incorporated with multivariable synthesis techniques to aid the design process. In particular, synthesis techniques used in an iterative procedure with the information furnished by heuristics and analysis tools should provide more powerful and efficient methodologies for multivariable feedback design.

Appendix A:

Properties of the Singular Value Decomposition

In this appendix we shall review some properties of the singular value decomposition. Our exposition is based upon that of[Ste73a]. Other useful references are [GoV83] and[KlL80]. The reader is referred to these sources for proofs and additional details. For brevity, we shall discuss only the case of square matrices; nonsquare matrices are discussed in the references.

Consider the matrix $A \in \mathbb{C}^{n \times n}$. There exist unitary matrices V and U such that A can be factored as

$$\begin{aligned} A &= V \Sigma U^H \\ &= \sum_{i=1}^{n} v_i u_i^H \sigma_i \end{aligned} \tag{A.1}$$

where $\Sigma = diag[\sigma_1, \sigma_2, \ldots, \sigma_n]$ and $\sigma_1 \geq \sigma_2 \geq \cdots \geq \sigma_n \geq 0$. The scalars σ_i are the *singular values* of A and the columns of $V = [v_1 | \cdots | v_n]$ and $U = [u_1 | \cdots | u_n]$ are, respectively, the left and right singular vectors. The singular values and vectors satisfy the identities

$$Au_i = \sigma_i v_i \tag{A.2}$$

$$v_i^H A = u_i^H \sigma_i \quad . \tag{A.3}$$

It can be shown that the singular values are the square roots of the eigenvalues of the matrices AA^H and $A^H A$. Furthermore, the left and right singular vectors are eigenvectors of AA^H and $A^H A$, respectively. Hence we have

$$A^H A U = U \Sigma^2 \tag{A.4}$$

and

$$AA^H V = V\Sigma^2 \quad . \tag{A.5}$$

The fact that singular values are the square roots of the eigenvalues of an Hermitian matrix implies that they possess many of the properties of such eigenvalues.

Although the singular values of A are uniquely defined, the singular vectors are not. If A has a singular value σ with multiplicity k, then the corresponding right singular vectors may be chosen as any orthonormal basis for the eigenspace of $A^H A$ corresponding to the eigenvalue σ^2. If $\sigma^2 > 0$ then, once the right singular vectors are chosen, the left singular vectors are uniquely determined from (A.2). Given a singular value with multiplity k, the associated k-dimensional eigenspaces of $A^H A$ and AA^H are uniquely defined and are termed the right and left singular subspaces associated with σ. These subspaces will be denoted U and V, respectively.

There even exists one degree of freedom in choosing a right singular vector corresponding to a singular value of multiplicity one. Let σ_i be such a singular value, let u_i and v_i be a corresponding pair of right and left singular vectors. Then, from (A.2), it follows that choosing $\hat{u}_i \triangleq e^{j\alpha}u_i$, $\alpha \in I\!R$, as a new right singular vector yields $\hat{v}_i = e^{j\alpha}v_i$ as the corresponding new left singular vector. Hence each pair of singular vectors corresponding to a distinct singular value is well-defined only to within multiplication by a unit magnitude scalar.

It is useful to denote the largest and smallest singular values of A by $\overline{\sigma} \triangleq \sigma_1$ and $\underline{\sigma} \triangleq \sigma_n$, respectively. The largest singular value of A is equal to the matrix norm of A induced by the standard Euclidean vector norm[44]

$$\bar{\sigma}[A] = ||A|| \quad . \tag{A.6}$$

If det$A \neq 0$, then the smallest singular value has the property that

$$\underline{\sigma}[A] = 1/||A^{-1}|| \quad . \tag{A.7}$$

The largest and smallest singular values of A provide upper and lower bounds on the magnitude of each eigenvalue of A

$$\bar{\sigma}[A] \geq |\lambda_i(A)| \geq \underline{\sigma}[A] \quad . \tag{A.8}$$

With no additional assumptions on A (e.g. $A = A^H$), there is no guarantee that the bound (A.8) is tight. One can construct examples wherein $\bar{\sigma}[A]$ is arbitrarily larger than the spectral radius of A. Likewise, $\underline{\sigma}[A]$ may be arbitrarily smaller than any of the eigenvalues of A. This is a significant observation, since it is well-known that $\underline{\sigma}[A]$ is a tight measure of the distance from A to the nearest singular matrix.

[44]In this monograph, $||\cdot||$ will refer to either the standard Euclidean vector norm or to the matrix norm induced by this vector norm.

Appendix B:

Properties of Differential Forms

The purpose of this appendix is to review some elementary properties of differential forms. Our review is not intended to be comprehensive; further details are found in many texts, e.g.[Spi65, Spi70], and[Ble81].

Consider the function $f: \mathbb{C} \rightarrow \mathbb{C}$ defined by $f(s) = g(x,y) + jh(x,y)$, where $s = x + jy$ and g and h are real-valued. Suppose that g and h have continuous partial derivatives, of all orders, with respect to the variables x and y. The *differential* of f is given by

$$df = \frac{\partial f}{\partial x} dx + \frac{\partial f}{\partial y} dy \tag{B.1}$$

The differential of f may be used to compute the derivative of f along a curve in the complex plane. Let $\gamma(t) = \gamma_1(t) + j\gamma_2(t)$, with $t \in [0,1]$ and $\gamma_i(t) \in I\!R$, be a smooth curve with tangent vector $\frac{d\gamma}{dt} = \frac{d\gamma_1}{dt} + j\,\frac{d\gamma_2}{dt}$. Consider the composite function $f \circ \gamma$: $[0,1] \rightarrow \mathbb{C}$. The derivative of f along γ is obtained by evaluating the differential form df on the tangent vector $\frac{\partial \gamma}{dt}$:

$$\begin{aligned} \frac{d(f \circ \gamma)}{dt} &= df\left(\frac{d\gamma}{dt}\right) \\ &= \frac{\partial f}{\partial x} dx\left(\frac{d\gamma}{dt}\right) + \frac{\partial f}{\partial y} dy\left(\frac{d\gamma}{dt}\right) \\ &= \frac{\partial f}{\partial x}\frac{d\gamma_1}{dt} + \frac{\partial f}{\partial y}\frac{d\gamma_2}{dt} \end{aligned} \tag{B.2}$$

The differential of a function is a special case of a more general construction known as a *differential 1-form*. One-forms are scalar-valued linear operators acting on vectors tangent to the complex plane (or some other differentiable manifold). Consider a one-form defined on the complex plane

$$P = Qdx + Rdy \quad , \tag{B.3}$$

where $Q : \mathbb{C} \to \mathbb{C}$ and $R : \mathbb{C} \to \mathbb{C}$ have continuous partial derivatives of all orders with respect to x and y. If there exists a function f (also termed a 0-form) such that $F = df$, then the 1-form F is said to be *exact*.

One can also define differential k-forms, which are scalar-valued multilinear functions of k tangent vectors. In addition to 0-forms and 1-forms, we shall need to consider 2-forms. Just as 1-forms appear in the integrands of line integrals, 2-forms appear in the integrands of surface integrals.

Given two 1-forms $P_i = Q_i dx + R_i dy$, the *wedge product* of P_1 and P_2 is the 2-form

$$\begin{aligned} P_1 \wedge P_2 &= (Q_1 dx + R_1 dy) \wedge (Q_2 dx + R_2 dy) \\ &= Q_1 R_2 dx \wedge dy + R_1 Q_2 dy \wedge dx \\ &= (Q_1 R_2 - Q_2 R_1) dx \wedge dy \end{aligned} \tag{B.4}$$

Note that the wedge product has the properties that $dx \wedge dy = -dy \wedge dx$ (hence $P_1 \wedge P_2 = -P_2 \wedge P_1$) and $dx \wedge dx = dy \wedge dy = 0$.

The operation d taking functions to 1-forms can be generalized to an operation taking k-forms to $k+1$-forms, provided that the necessary differentiability properties are present. Given the 1-form (B.3), the 2-form dP is equal to

$$\begin{aligned} dP &= dQ \wedge dx + dR \wedge dy \\ &= \left[\frac{\partial Q}{\partial x} dx + \frac{\partial Q}{\partial y} dy \right] \wedge dx + \left[\frac{\partial R}{\partial x} dx + \frac{\partial R}{\partial y} dy \right] \wedge dy \\ &= \left[\frac{\partial R}{\partial x} - \frac{\partial Q}{\partial y} \right] dx \wedge dy \quad . \end{aligned} \tag{B.5}$$

A 1-form P with the property that $dP = 0$ is called *closed*. It turns out that every exact form is closed. To show that this is true, suppose that $P = df$, where f is a function with continuous partial derivatives of all orders. Then $\frac{\partial^2 f}{\partial y \partial x} = \frac{\partial^2 f}{\partial x \partial y}$, and it follows from (B.5) that

$dP = d^2f = 0$. The Poincaré Lemma ([Spi65], p.94) provides a converse to this statement. If $D \subsetneq \mathbb{C}$ is a convex open set, then every closed form on D is also exact. Hence, given any 1-form P with the property that $dP = 0$, we can find a function f such that $P = df$.

Differential forms can, of course, be defined on differentiable manifolds other than the complex plane. It is also possible to define forms taking values in a vector space such as the Lie algebra of a Lie group[Ble81].

We shall have need to use Stokes' Theorem applied to integration in the complex plane. Suppose that $P = Qdx + Rdy$ is a differential 1-form defined on a simply-connected open subset of the complex plane. Let C be a simple closed curve lying in this subset, and denote the interior of C by D. Then Stoke's Theorem states that

$$\int_C P = \int_D dP \quad , \tag{B.6}$$

where the line integral is taken in the counterclockwise direction and the 2-form dP is given by (B.5). Note that if P is a *closed* form then $\int_C P = 0$. It follows easily that integration of a closed 1-form is *path-independent*.

Suppose that $P = fds$. Then

$$\begin{aligned} dP &= df \wedge ds \\ &= j \left[\frac{\partial f}{\partial x} + j \frac{\partial f}{\partial y} \right] dx \wedge dy \end{aligned} \tag{B.7}$$

and it follows that P is closed if and only if the Cauchy-Riemann equations (8.2.1) are satisfied.

The standard basis vectors for the tangent space to $\mathbb{C}$ are commonly denoted $\{ \frac{\partial}{\partial x} , \frac{\partial}{\partial y} \}$.

Another useful basis is given by the set $\{ \frac{\partial}{\partial s} , \frac{\partial}{\partial \bar{s}} \}$ defined by [GrH78]

$$\frac{\partial}{\partial s} \triangleq \frac{1}{2}\left[\frac{\partial}{\partial x} - j\frac{\partial}{\partial y}\right]$$
$$\frac{\partial}{\partial \bar{s}} \triangleq \frac{1}{2}\left[\frac{\partial}{\partial x} + j\frac{\partial}{\partial y}\right] . \tag{B.8}$$

Note that the function $f(s) = g(s) + jh(s)$ is complex analytic if and only if

$$\frac{\partial f}{\partial \bar{s}} = \frac{1}{2}\left[\frac{\partial f}{\partial x} + j\frac{\partial f}{\partial y}\right]$$
$$= 0 . \tag{B.9}$$

It is easy to verify that the dual basis for the cotangent space is [GrH78]

$$ds = dx + jdy$$
$$d\bar{s} = dx - jdy . \tag{B.10}$$

Note that

$$ds \wedge d\bar{s} = -2j dx \wedge dy \tag{B.11}$$

Expressed in these coordinates, Stokes Theorem, when applied to the 1-form $P = f ds$, yields

$$\int_C f ds = -\int_D \frac{\partial f}{\partial \bar{s}} ds \wedge d\bar{s} , \tag{B.12}$$

where the integration around C is taken in the counter-clockwise direction and the minus sign appears because the bases $\{dx, dy\}$ and $\{ds, d\bar{s}\}$ have opposite orientation.

Finally, we shall need one additional operation on differential forms: the star operator[Ble81]. For our purposes, it suffices to define this operator on real-valued 1-forms. Given the 1-form $P = Qdx + Rdy$, where Q and R are now assumed to take *real* values, define

$$*P \triangleq -Rdx + Qdy . \tag{B.13}$$

Note that if $P = df$, where f is real-valued, then

$$P \wedge *P = \left[\left(\frac{\partial f}{\partial x} \right)^2 + \left(\frac{\partial f}{\partial y} \right)^2 \right] dx \wedge dy \quad . \tag{B.14}$$

Note also that

$$d*df = \left(\frac{\partial^2 f}{\partial x^2} + \frac{\partial^2 f}{\partial y^2} \right) dx \wedge dy \quad . \tag{B.15}$$

Hence, if f is *harmonic*, then $d*df = 0$.

Appendix C:

Proofs of Theorems in Chapter 5

Proof of Theorem 5.3.1: The assumption of closed loop stability implies that the sensitivity function $S_O(s)$ has no poles in the closed right half plane. Hence $\det S_O(s)$ has no poles in the closed right half plane. Furthermore, $\det S_O(s)$ has zeros at the open right half plane poles of $L_O(s)$. Removing these zeros with a Blaschke product

$$B_p(s) = \prod_{i=1}^{N_p} \frac{p_i - s}{\overline{p}_i + s} \tag{C.1}$$

yields a proper rational function

$$D(s) \triangleq B_p^{-1}(s) \cdot \det S_O(s) \tag{C.2}$$

with no poles or zeros in the open right half plane. Note also that $\log|D(j\omega)| = \log|\det S_O(j\omega)|$. Straightforward analysis reveals that

$$\begin{aligned} \log \det S_O(s) &= -\log \det[I + L_O(s)] \\ &= -\log[1 + \sum_k L_k(s)] \end{aligned} \tag{C.3}$$

where the functions $L_k(s)$ have at least one factor $L_{ij}(s)$ and hence have at least two more poles than zeros. Since the term ΣL_k approaches zero at infinity, there exists a frequency ω_1 such that $|\Sigma L_k(j\omega)| < 1, \forall\ \omega \geq \omega_1$. At these frequencies (C.3) has the expansion [Kno56]

$$\log\det S_O = -\sum_k L_k + \frac{(\sum_k L_k)^2}{2} + \text{higher order terms} \quad . \tag{C.4}$$

Thus $\omega \log|\det S_O(j\omega)| \rightarrow 0$ as $\omega \rightarrow \infty$. At this point, the proof of Theorem 3.3.1 (see the Appendix of[FrL85a]) may followed to obtain the desired result.

■

Appendix D:

Proof of Theorem 10.2.3

We need the following lemma, whose proof is a straightforward calculation.

Lemma D.1: Consider the differential form $f = adx + bdy$, where $a \in \mathbb{C}$ and $b \in \mathbb{C}$ Then

$$\bar{f} \wedge f = 2j \,\mathrm{Im}[\bar{a}b]dx \wedge dy$$

and (D.1)

$$f \wedge \bar{f} = -2j \,\mathrm{Im}[\bar{a}b]dx \wedge dy$$

Furthermore, if a and b are both *real* valued, then

$$\bar{f} \wedge f = f \wedge \bar{f} = 0 \quad . \tag{D.2}$$

Similarly, if a and b are both take *imaginary* values, then

$$\bar{f} \wedge f = f \wedge \bar{f} = 0 \quad . \tag{D.3}$$

∎

Define

$$A \triangleq j[-\frac{\partial \log \sigma_i}{\partial y} dx + \frac{\partial \log \sigma_i}{\partial x} dy] \tag{D.4}$$

and note the proof of Corollary 10.2.2. implies that

$$A = v_i^H dv_i - u_i^H du_i \quad . \tag{D.5}$$

Hence

$$\begin{aligned} dA &= j(\nabla^2 \log \sigma_i) dx \wedge dy \\ &= dv_i^H \wedge dv_i - du_i^H \wedge du_i \end{aligned} \tag{D.6}$$

where we used the fact that the exact forms dv_i and du_i are closed (Appendix B). From

(10.2.1) we have

$$du_i^H \wedge du_i = \sum_{k=1}^{n} (du_i^H u_k) \wedge (u_k^H du_i)$$

and (D.7)

$$dv_i^H \wedge dv_i = \sum_{k=1}^{n} (dv_i^H v_k) \wedge (v_k^H dv_i)$$

so that

$$dA = \sum_{k=1}^{n} (dv_i^H v_k) \wedge (v_k^H dv_i) - \sum_{k=1}^{n} (du_i^H u_k) \wedge (u_k^H du_i) \tag{D.8}$$

Using (10.2.2) yields

$$(du_i^H u_k) \wedge (u_k^H du_i) =$$

$$\left[\frac{1}{\sigma_i^2 - \sigma_k^2}\right]^2 \Big\{ \sigma_i^2 (v_i^H dMu_k) \wedge (v_i^H dMu_k)^H + \sigma_k^2 (v_k^H dMu_i)^H \wedge (v_k^H dMu_i) \tag{D.9}$$

$$+ \sigma_i \sigma_k [(v_i^H dMu_k) \wedge (v_k^H dMu_i) + (v_k^H dMu_i)^H \wedge (v_i^H dMu_k)^H] \Big\}$$

and

$$(dv_i^H v_k) \wedge (v_k^H dv_i) =$$

$$\left[\frac{1}{\sigma_i^2 - \sigma_k^2}\right]^2 \Big\{ \sigma_i^2 (v_k^H dMu_i)^H \wedge (v_k^H dMu_i) + \sigma_k^2 (v_i^H dMu_k) \wedge (v_i^H dMu_k)^H \tag{D.10}$$

$$+ \sigma_i \sigma_k [(v_i^H dMu_k)^H \wedge (v_k^H dMu_k)^H + (v_i^H dMu_k) \wedge (v_k^H dMu_i)] \Big\}$$

Substituting (D.9) and (D.10) into (D.8) yields

$$dA = \sum_{\substack{k=1 \\ k\neq i}} \left[\frac{1}{\sigma_i^2 - \sigma_k^2}\right] \left[-(v_i^H dMu_k)\wedge(v_i^H dMu_k)^H + (v_k^H dMu_i)^H \wedge (v_k^H dMu_i)\right]$$

$$+ (dv_i^H v_i)\wedge(v_i^H dv_i) - (du_i^H u_i)\wedge(u_i^H du_i) \quad . \tag{D.11}$$

Since $v_i^H dv_i$ and $u_i^H du_i$ take pure imaginary values, (D.3) shows that the last two terms in (D.11) equal zero.

To simplify the remaining terms, first note that (D.1) implies

$$(v_i^H dMu_k)\wedge(v_i^H dMu_k)^H = -2j \,\mathrm{Im}\left[(v_i^H \frac{\partial M}{\partial x} u_k)^H (v_i^H \frac{\partial M}{\partial y} u_k)\right] dx\wedge dy$$

and (D.12)

$$(v_k^H dMu_i)^H \wedge (v_k^H dMu_i) = 2j \,\mathrm{Im}\left[(v_k^H \frac{\partial M}{\partial x} u_i)^H (v_k^H \frac{\partial M}{\partial y} u_i)\right] dx\wedge dy \quad .$$

Using the fact (Lemma 8.3.2) that

$$v_k^H \frac{\partial M}{\partial x} u_i = \frac{1}{j} v_k^H \frac{\partial M}{\partial y} u_i \tag{D.13}$$

yields

$$\begin{aligned}(v_i^H dMu_k)\wedge(v_i^H dMu_k)^H &= -2j\,\mathrm{Im}[j\,(v_i^H \frac{\partial M}{\partial y} u_k)^H (v_i^H \frac{\partial M}{\partial y} u_k)]\, dx\wedge dy \\ &= -2j\,|v_i^H \frac{\partial M}{\partial y} u_k|^2\, dx\wedge dy \\ &= -j \left[|v_i^H \frac{\partial M}{\partial x} u_k|^2 + |v_i^H \frac{\partial M}{\partial y} u_k|^2\right] dx\wedge dy\end{aligned} \tag{D.14}$$

Similarly

$$(v_k^H dMu_i)^H \wedge (v_k^H dMu_i) = j \left[|v_k^H \frac{\partial M}{\partial x} u_i|^2 + |v_k^H \frac{\partial M}{\partial y} u_i|^2\right] dx\wedge dy \tag{D.15}$$

Using (D.14) and (D.15) in (D.11) yields the desired result.

Appendix E:

Proof of Theorem 11.2.1

Proof of Theorem 11.2.1: Consider the closed curve $C(R,\varepsilon)$ pictured in Figure E.1. Denote the interior of $C(R,\varepsilon)$ by $D(R,\varepsilon)$, the large semicircle by C_R, the small semicircle by C_ε, and let C_1 and C_2 be the segments of the $j\omega$-axis shown in the figure. Define

$$f(s) \triangleq \log\sigma(s) + j[\delta\theta_w(u,v)(s) - \delta\theta_w(u,v)(0)] \tag{E.1}$$

To simplify the notation used in the proof, we shall replace ω_0 and ω in the statement of the theorem by ω and y, respectively. Then Stokes' Theorem (Appendix B) yields

$$\int_{C(R,\varepsilon)} \frac{f(s)}{s - j\omega}\, ds = - \int_{D(R,\varepsilon)} \frac{\partial f}{\partial \bar{s}} \left[\frac{1}{s - j\omega} \right] ds \wedge d\bar{s} \ . \tag{E.2}$$

The remainder of the proof will be broken into four major steps. The first step will be to evaluate the limit as $\varepsilon \to 0$ in (E.2). Next, we shall evaluate the limit as $R \to \infty$ of this result. Third, conjugate symmetry will be invoked to simplify the resulting integral. Finally, an argument using changes of variable and integration by parts will yield the desired result.

Step 1: (take the limit as $\varepsilon \to 0$ in (E.2)) We first show that the limit

$$\int_{C(R,0)} \frac{f(s)}{s - j\omega}\, ds \triangleq \lim_{\varepsilon \to 0} \int_{C(R,\varepsilon)} \frac{f(s)}{s - j\omega}\, ds \tag{E.3}$$

exists and is finite; we shall do this by showing that the limit as $\varepsilon \to 0$ of the right hand side of (E.2) has these properties. Define a polar coordinate system on the right half plane by

$$\begin{aligned} s &= x + jy \\ &= j\omega + re^{j\alpha}\,, \quad r \in [0,\infty)\,, \quad \alpha \in [-\pi/2, \pi/2] \end{aligned} \tag{E.4}$$

Since $ds = dre^{j\alpha} + re^{j\alpha} j d\alpha$ and $d\bar{s} = dre^{-j\alpha} - re^{-j\alpha} j d\alpha$, it follows readily that

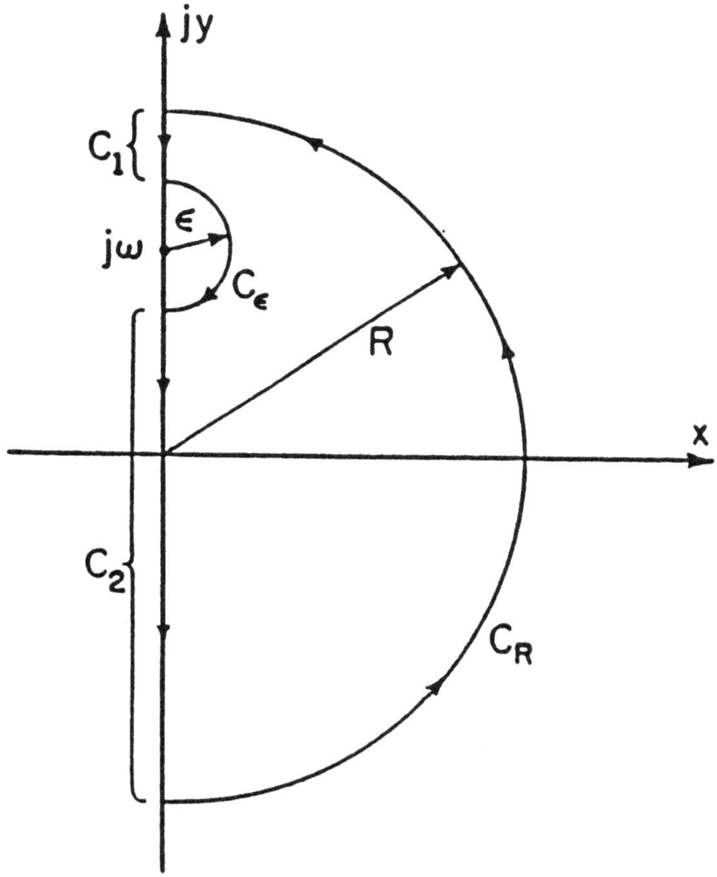

Figure E.1: The Curve $C(R,\varepsilon)$

$ds \wedge d\bar{s} = -2jrdr \wedge d\alpha$. Hence, in polar coordinates, the right hand side of (E.2) becomes

$$-\int_{D(R,\varepsilon)} \frac{\partial f}{\partial \bar{s}} \left[\frac{1}{s - j\omega}\right] ds \wedge d\bar{s} = 2j \int_{D(R,\varepsilon)} \frac{\partial f}{\partial \bar{s}} \left[e^{-j\alpha}\right] dr \wedge d\alpha \tag{E.5}$$

Since the integrand of (E.5) is independent of ε, it follows that the limit

$$\int_{D(R,0)} \frac{\partial f}{\partial \bar{s}} \left[e^{-j\alpha}\right] dr \wedge d\alpha \triangleq \lim_{\varepsilon \to 0} \int_{D(R,\varepsilon)} \frac{\partial f}{\partial \bar{s}} \left[e^{-j\alpha}\right] dr \wedge d\alpha \tag{E.6}$$

exists and is finite. Hence (E.3) also has these properties.

Using (E.3) and (E.6) in (E.2) yields

$$\int_{C(R,0)} \frac{f(s)}{s-j\omega}\, ds = -\int_{D(R,0)} \frac{\partial f}{\partial \bar{s}} \left[\frac{1}{s-j\omega}\right] ds \wedge d\bar{s} \ . \tag{E.7}$$

Next, we decompose the left hand side of (E.7) according to the decomposition of the curve $C(R,\varepsilon)$ in Figure E.1:

$$\begin{aligned}\int_{C(R,0)} \frac{f(s)}{s-j\omega}\, ds &= \lim_{\varepsilon \to 0} \int_{C_\varepsilon} \frac{f(s)}{s-j\omega}\, ds + \int_{C_R} \frac{f(s)}{s-j\omega}\, ds \\ &+ \lim_{\varepsilon \to 0} \left[\int_{R}^{\omega+\varepsilon} f(jy) \frac{dy}{y-\omega} + \int_{\omega-\varepsilon}^{-R} f(jy) \frac{dy}{y-\omega} \right]\end{aligned} \tag{E.8}$$

Consider the first integral on the right hand side of (E.8). Parameterize the curve C_ε by $s(t) = j\omega + \varepsilon \exp[j(\pi/2 - \pi t)]$, $0 \le t \le 1$. Then $ds = \varepsilon \exp[j(\pi/2 - \pi t)](-j\pi)dt$ and continuity implies that

$$\begin{aligned}\lim_{\varepsilon \to 0} \int_{C_\varepsilon} \frac{f(s)}{s - j\omega}\, ds &= -j\pi \int_0^1 f(j\omega + \varepsilon\exp[j(\pi/2 - \pi t)])dt \\ &= -j\pi f(j\omega) \ .\end{aligned} \tag{E.9}$$

It follows that the second limit on the right hand side of (E.8) exists and is finite; hence, define

$$\int_R^{-R} f(jy)\,\frac{dy}{y-\omega} \triangleq \lim_{\varepsilon\to 0}\left[\int_R^{\omega+\varepsilon} f(jy)\,\frac{dy}{y-\omega} + \int_{\omega-\varepsilon}^{-R} f(jy)\,\frac{dy}{y-\omega}\right] \quad . \tag{E.10}$$

Substituting (E.9) and (E.10) into (E.8) yields

$$\int_{C(R,0)} \frac{f(s)}{s-j\omega}\,ds = -j\pi f(j\omega) + \int_{C_R} \frac{f(s)}{s-j\omega}\,ds + \int_R^{-R} f(jy)\frac{dy}{y-\omega} \tag{E.11}$$

Substituting (E.11) into (E.7) and rearranging, we have

$$\begin{aligned} f(j\omega) = \frac{1}{j\pi}\int_{C_R} f(s)\left[\frac{1}{s-j\omega}\right]ds + \frac{1}{j\pi}\int_R^{-R} f(jy)\frac{dy}{y-\omega} \\ + \frac{1}{j\pi}\int_{D(R,0)} \frac{\partial f}{\partial \bar{s}}\left[\frac{1}{s-j\omega}\right]ds\wedge d\bar{s} \quad . \end{aligned} \tag{E.12}$$

Similar calculations show that

$$\begin{aligned} f(-j\omega) = \frac{1}{j\pi}\int_{C_R} f(s)\left[\frac{1}{s+j\omega}\right]ds + \frac{1}{j\pi}\int_R^{-R} f(jy)\,\frac{dy}{y+\omega} \\ + \frac{1}{j\pi}\int_{D(R,0)} \frac{\partial f}{\partial \bar{s}}\left[\frac{1}{s+j\omega}\right]ds\wedge d\bar{s} \quad . \end{aligned} \tag{E.13}$$

Subtracting (E.13) from (E.12) and using conjugate symmetry yields

$$\begin{aligned} \delta\theta_w(u,v)(j\omega) - \delta\theta_w(u,v)(0) = &-\frac{1}{2\pi}\int_{C_R} f(s)\left[\frac{1}{s-j\omega} - \frac{1}{s+j\omega}\right]ds \\ &-\frac{1}{2\pi}\int_R^{-R} f(jy)\left[\frac{1}{y-\omega} - \frac{1}{y+\omega}\right]dy \\ &-\frac{1}{2\pi}\int_{D(R,0)} \frac{\partial f}{\partial \bar{s}}\left[\frac{1}{s-j\omega}\right]ds\wedge d\bar{s} \\ &+\frac{1}{2\pi}\int_{D(R,0)} \frac{\partial f}{\partial \bar{s}}\left[\frac{1}{s+j\omega}\right]ds\wedge d\bar{s} \end{aligned} \tag{E.14}$$

Step 2: (take the limit as $R \to \infty$) We shall first show that the following two limits exist and are finite:

$$\begin{aligned} L_1 &\triangleq \lim_{R\to\infty} \int_{C_R} f(s)\left[\frac{1}{s-j\omega} - \frac{1}{s+j\omega}\right] ds \\ &= \lim_{R\to\infty} \int_{C_R} f(s)\left[\frac{2j\omega}{s^2+\omega^2}\right] ds \end{aligned} \tag{E.15}$$

and

$$\begin{aligned} L_2 &\triangleq \lim_{R\to\infty} \int_{R}^{-R} f(jy)\left[\frac{1}{y-\omega} - \frac{1}{y+\omega}\right] dy \\ &= \lim_{R\to\infty} \int_{-R}^{R} f(jy)\left[\frac{2\omega}{\omega^2-y^2}\right] dy \quad . \end{aligned} \tag{E.16}$$

We shall also show that the first limit equals zero.

Define a polar coordinate system centered at the origin: $s = Re^{j\alpha}$, $R \in [0,\infty)$, $\alpha \in [-\frac{\pi}{2},\frac{\pi}{2}]$. In these coordinates

$$L_1 = \lim_{R\to\infty} -2\omega \int_{-\frac{\pi}{2}}^{\frac{\pi}{2}} f(Re^{j\alpha})\,[\,Re^{j\alpha}/(R^2e^{2j\alpha}+\omega^2)]d\alpha \tag{E.17}$$

Define

$$M(R) \triangleq \sup_{\alpha\in[-\pi/2,\pi/2]} \left| f(Re^{j\alpha}) \left[Re^{j\alpha}/(R^2e^{2j\alpha}+\omega^2) \right] \right| \tag{E.18}$$

Assuming with no loss of generality that $R > \omega$, notice that

$$M(R) \le \sup_{\alpha\in[-\pi/2,\pi/2]} \left| f(Re^{j\alpha}) \right| \frac{R}{R^2-\omega^2} \tag{E.19}$$

We need to show that $\lim_{R\to\infty} M(R) = 0$. Since $f = \log\{\sigma \exp[j\delta\theta_w(u,v)]\}$ this fact follows from assumption (vi) of Theorem 11.2.1 applied to the right hand side of (E.19).

Combining (E.17-19), we have

$$|L_1| \leq \lim_{R \to \infty} 2\omega \int_{C_R} |M(R)|\, d\alpha \quad , \tag{E.20}$$

and it follows that $L_1 = 0$. Arguments similar to these may be used to show that the limit L_2 in (E.16) exists and is finite. (The integrand in (E.16) approaches zero as $y \to \infty$ faster than $1/y \to 0$; the result follows from this observation and a standard argument.) The fact that $f(jy)$ is conjugate symmetric yields

$$L_2 = -\, 4\omega \int_0^{\infty} \log\sigma(jy) \left[\frac{1}{y^2 - \omega^2} \right] dy \tag{E.21}$$

Taking the limit as $R \to \infty$ in (E.14), and using (E.21) and the fact that $L_1 = 0$, it follows that the limit

$$\begin{aligned} L_3 \triangleq \lim_{R \to \infty} \Bigg\{ & - \int_{D(R,0)} \frac{\partial f}{\partial \bar{s}} \left[\frac{1}{s - j\omega} \right] ds \wedge d\bar{s} \\ & + \int_{D(R,0)} \frac{\partial f}{\partial \bar{s}} \left[\frac{1}{s + j\omega} \right] ds \wedge d\bar{s} \Bigg\} \quad . \end{aligned} \tag{E.22}$$

exists and is finite. Hence, from (E.14)

$$\delta\theta_w(u,v)(j\omega) - \delta\theta_w(u,v)(0) = \frac{2\omega}{\pi} \int_0^{\infty} \log\sigma \left[\frac{1}{y^2 - \omega^2} \right] dy + \frac{1}{2\pi} L_3 \quad . \tag{E.23}$$

Step 3: (simplifications using conjugate symmetry) We shall simplify the limit (E.22). Since the limit as $R \to \infty$ of the region $D(R,0)$ is equal to the closed right half plane, we shall denote (E.22) by

$$L_3 = - \int_{CRHP} \frac{\partial f}{\partial \bar{s}} \left[\frac{1}{s-j\omega} \right] ds \wedge d\bar{s} + \int_{CRHP} \frac{\partial f}{\partial \bar{s}} \left[\frac{1}{s+j\omega} \right] ds \wedge d\bar{s} \quad . \tag{E.24}$$

Define

$$I_1 = \int_{CRHP} \frac{\partial f}{\partial \bar{s}} \left[\frac{1}{s-j\omega} \right] ds \wedge d\bar{s} \tag{E.25}$$

and

$$I_2 = \int_{CRHP} \frac{\partial f}{\partial \bar{s}} \left[\frac{1}{s+j\omega} \right] ds \wedge d\bar{s} \tag{E.26}$$

By invoking conjugate symmetry, we shall show that

$$I_1 = - \bar{I}_2 \quad . \tag{E.27}$$

First, we need a simple lemma.

Let $g(s)$ and $h(s)$ be real-valued functions of the complex variable $s = x + jy$ and define the complex-valued function $f(s) = g(s) + jh(s)$. When we identify the complex plane with $\mathbb{R}^2$, we shall also use the notation $f(x,y) = g(x,y) + jh(x,y)$. Recall that the function $f(x,y)$ is *conjugate symmetric* if $f(x,y) = \overline{f(x,-y)}$ (equivalently, if $g(x,y) = g(x,-y)$ and $h(x,y) = -h(x,-y)$).

Lemma E.1: Suppose that the partial derivatives of g and h with respect to x and y exist and are continuous. Then

$$\frac{\partial g}{\partial x}(x,y) = \frac{\partial g}{\partial x}(x,-y) \tag{E.28a}$$

$$\frac{\partial g}{\partial y}(x,y) = - \frac{\partial g}{\partial y}(x,-y) \tag{E.28b}$$

$$\frac{\partial h}{\partial x}(x,y) = \frac{\partial h}{\partial x}(x,-y) \tag{E.28c}$$

$$\frac{\partial h}{\partial y}(x,y) = -\frac{\partial h}{\partial y}(x,-y) \quad . \tag{E.28d}$$

■

Proof: By definition

$$\begin{aligned}\frac{\partial g}{\partial y}(x,y) &= \lim_{\Delta y \to 0} \frac{g(x,y+\Delta y) - g(x,y)}{\Delta y} \\ &= \lim_{\Delta y \to 0} \frac{g(x,-y-\Delta y) - g(x,-y)}{\Delta y}\end{aligned}$$

Define $\Delta\hat{y} \triangleq -\Delta y$. Then

$$\begin{aligned}\frac{\partial g}{\partial y}(x,y) &= -\lim_{\Delta\hat{y} \to 0} \frac{g(x,-y+\Delta\hat{y}) - g(x,-y)}{\Delta\hat{y}} \\ &= -\frac{\partial g}{\partial y}(x,-y)\end{aligned}$$

Proofs of the other identities are similar . ■

From Lemma E.1, it follows that the function

$$\frac{\partial f}{\partial \bar{s}} \triangleq \frac{1}{2}\left[\frac{\partial f}{\partial x} + j\frac{\partial f}{\partial y}\right] , \tag{E.29}$$

defined in Appendix B, is conjugate symmetric:

$$\frac{\partial f}{\partial \bar{s}}(x,y) = \overline{\frac{\partial f}{\partial \bar{s}}(x,-y)} \quad . \tag{E.30}$$

Since (Appendix B), $ds \wedge d\bar{s} = -2j dx \wedge dy$, (E.25) reduces to

$$I_1 = -2j \int_{-\infty}^{\infty} \int_{0}^{\infty} \frac{\partial f}{\partial \bar{s}}(x,y) \left[\frac{1}{x+j(y-\omega)} \right] dx\, dy$$

$$= -2j \int_{0}^{\infty} \int_{0}^{\infty} \frac{\partial f}{\partial \bar{s}}(x,y) \left[\frac{1}{x+j(y-\omega)} \right] dx\, dy \tag{E.31}$$

$$-2j \int_{-\infty}^{0} \int_{0}^{\infty} \frac{\partial f}{\partial \bar{s}}(x,y) \left[\frac{1}{x+j(y-\omega)} \right] dx\, dy$$

Making the change of variable $\hat{y} \triangleq -y$, and using (E.30), the second integral in (E.31) becomes

$$-2j \int_{0}^{\infty} \int_{0}^{\infty} \overline{\frac{\partial f}{\partial \bar{s}}(x,\hat{y})} \left[\frac{1}{x-j(\hat{y}+\omega)} \right] dx\, d\hat{y} \quad . \tag{E.32}$$

Since $\hat{y}$ is just a dummy variable, we can replace y by y in (E.32), and substitute into (E.31) to show that

$$I_1 = -2j \int_{0}^{\infty} \int_{0}^{\infty} \left\{ \frac{\partial f}{\partial \bar{s}}(x,y) \left[\frac{1}{x+j(y-\omega)} \right] + \overline{\frac{\partial f}{\partial \bar{s}}(x,y)} \left[\frac{1}{x - j(y + \omega)} \right] \right\} dx\, dy \quad . \tag{E.33}$$

A similar computation yields

$$I_2 = -2j \int_{0}^{\infty} \int_{0}^{\infty} \left\{ \frac{\partial f}{\partial \bar{s}}(x,y) \left[\frac{1}{x+j(y+\omega)} \right] + \overline{\frac{\partial f}{\partial \bar{s}}(x,y)} \left[\frac{1}{x-j(y-\omega)} \right] \right\} dx\, dy \quad . \tag{E.34}$$

Together, equations (E.33) and (E.34) yield (E.27).

We next simplify (E.24) by computing

$$\begin{aligned} L_3 &= -I_1 + I_2 \\ &= -2\mathrm{Re}[I_1] \\ &= -2\mathrm{Re} \int_{CRHP} \frac{\partial f}{\partial \bar{s}} \left[\frac{1}{s - j\omega} \right] ds \wedge d\bar{s} \end{aligned} \tag{E.35}$$

For notational brevity, define $\theta \triangleq \delta\theta_w(u,v)$. Then, expressed in the polar coordinates (E.4), the terms in the integrand of (E.35) become

$$\frac{\partial f}{\partial \bar{s}} = \frac{1}{2} e^{j\alpha} \left[\left[\frac{\partial \log\sigma}{\partial r} - \frac{1}{r} \frac{\partial \theta}{\partial \alpha} \right] + j \left[\frac{\partial \theta}{\partial r} + \frac{1}{r} \frac{\partial \log\sigma}{\partial \alpha} \right] \right] , \tag{E.36}$$

$$\frac{1}{s - j\omega} = \frac{1}{r} e^{-j\alpha} , \tag{E.37}$$

and

$$ds \wedge d\bar{s} = -2jr dr \wedge d\alpha . \tag{E.38}$$

Substituting these into (E.35) yields

$$\begin{aligned} L_3 &= 2\mathrm{Re} \int_{CRHP} \left[j \left[\frac{\partial \log\sigma}{\partial r} - \frac{1}{r} \frac{\partial \theta}{\partial \alpha} \right] - \left[\frac{\partial \theta}{\partial r} + \frac{1}{r} \frac{\partial \log\sigma}{\partial \alpha} \right] \right] dr \wedge d\alpha \\ &= -2 \int_{CRHP} \left[\frac{\partial \theta}{\partial r} + \frac{1}{r} \frac{\partial \log\sigma}{\partial \alpha} \right] dr \wedge d\alpha \\ &= -2 \int_{-\frac{\pi}{2}}^{\frac{\pi}{2}} \int_0^{\infty} \left[\frac{\partial \theta}{\partial r} + \frac{1}{r} \frac{\partial \log\sigma}{\partial \alpha} \right] dr \, d\alpha \end{aligned} \tag{E.39}$$

Using (E.39) in (E.23) yields, finally,

$$\begin{aligned} \delta\theta_w(u,v)(j\omega) - \delta\theta_w(u,v)(0) = \; & \frac{2\omega}{\pi} \int_0^{\infty} \log\sigma \left[\frac{1}{y^2 - \omega^2} \right] dy \\ & - \frac{1}{\pi} \int_{-\frac{\pi}{2}}^{\frac{\pi}{2}} \int_0^{\infty} \left[\frac{\partial \delta\theta_w(u,r)}{\partial r} + \frac{1}{r} \frac{\partial \log\sigma}{\partial \alpha} \right] dv \, d\alpha \end{aligned} \tag{E.40}$$

Step 4: The derivation of the scalar gain-phase relation [Bod45] may be followed to show that the first integral in (E.40) is equal to the first integral in (11.2.6). Equations (10.3.19), when transformed into polar coordinates, yield

$$\frac{1}{r}\,\frac{\partial \log\sigma}{\partial\alpha} + \frac{\partial\delta\theta_w(u,v)}{\partial r} = \frac{-1}{j}\left[\rho_w^H\,\frac{\partial[\rho_w(\mathbf{V})]}{\partial r} - \rho_w^H\,\frac{\partial[\rho_w(\mathbf{U})]}{\partial r}\right] . \qquad \text{(E.41)}$$

Substituting (E.41) into (E.40) and then (11.2.6) yields the desired result.

■

References

[ALP83] Final Technical Report, "Robustness analysis of beam pointing control systems," TR-168-1, ALPHATECH, Inc., Burlington, MA, May 1983.

[And68] B.D.O. Anderson, "A simplified viewpoint of hyperstability," *IEEE Trans. on Auto. Cont.*, June 1968.

[AsH84] K.J. Astrom and T. Hagglund, "A frequency domain method for automatic tuning of simple feedback loops," *Proc. 23rd IEEE Conf. on Decision and Contr.*, Dec. 1984.

[AuM77] L. Auslander and R.E. MacKenzie, *Introduction to Differentiable Manifolds*, NY: Dover, 1977.

[BjG73] A. Bjorck and G.H. Golub, "Numerical methods for computing angles between linear subspaces," *Mathematics of Computation*, vol. 27, no. 123, pp. 579-594, July 1973.

[Ble81] D. Bleecker, *Gauge Theory and Variational Principles*, London: Addison-Wesley, 1981.

[Bod45] H.W. Bode, *Network Analysis and Feedback Amplifier Design*, Princeton, NJ: Van Nostrand, 1945.

[BoD84] S. Boyd and C.A. Desoer, "Subharmonic functions and performance bounds on linear time-invariant feedback systems," UCB/ERL M84/51, Tech. Rept., Univ. of California, Berkeley, CA, June 1984. Also, *IMA Jour. of Infor. and Contr., 1985*.

[Boy86] S.P. Boyd, "A new CAD method and associated architectures for linear controllers," *Infor. Systems Lab*, Tech. Rept. No. L-104-86-1, Stanford Univ., Dec. 1986.

[CaD82] F.M. Callier and C.A. Desoer, *Multivariable Feedback Systems*, NY: Springer-Verlag, 1982.

[CFL81] J.B. Cruz, Jr., J.S. Freudenberg, and D.P. Looze, "A relation between sensitivity and stability of multivariable feedback systems," *IEEE Trans. Auto Contr.*, vol. AC-26, pp. 66-74, Feb. 1981.

[Con78] J.B. Conway, *Functions of One Complex Variable*, 2nd ed., NY: Springer-Verlag, 1978.

[CrP64] J.B. Cruz, Jr., and W.R. Perkins, "A new approach to the sensitivity problem in multivariable feedback system design," *IEEE Trans. Auto. Contr.*, vol. AC-9, pp. 210-223, 1964.

[DeS74] C.A. Desoer and J.D. Schulman, "Poles and zeros of matrix transfer functions and their dynamical interpretation," *IEEE Trans. on Circuits and Systems*, vol. CAS-21, no. 1, Jan. 1974.

[DoS81] J.C. Doyle and G. Stein, "Multivariable feedback design: concepts for a classical/modern synthesis," *IEEE Trans on Auto. Contr.*, vol. AC-26, Feb. 1981.

[Doy79] J.C. Doyle, "Robustness of multiloop linear feedback systems," *Proc. 17th IEEE Conf. on Decision and Contr.*, Jan. 1979.

[Doy82] J.C. Doyle, "Analysis of feedback systems with structured uncertainties," *IEEE Proc.*, vol. 129, Pt.D, no. 6, pp. 1-10, Nov. 1982.

[Doy85] J.C. Doyle, "Structured uncertainty in control system design," *Proc. 1985 Conf. on Decision and Contr.*, Dec. 1985.

[DWS82] J.C. Doyle, J.E. Wall, Jr., and G. Stein, "Performance and robustness analysis for structured uncertainty," *Proc. 21st IEEE Conf. on Decision and Contr.*, pp. 629-636, Orlando, FL, Dec. 1982.

[FaK80] H.M. Farkas and I. Kra, *Riemann Surfaces*, NY: Springer-Verlag, 1980.

[FLC82] J.S. Freudenberg, D.P. Looze and J.B. Cruz, Jr., "Robustness analysis using singular value sensitivities," *Int. J. Contr.*, vol. 35, no. 1, p. 95, 1982.

[FPE86] G.F. Franklin, J.D. Powell and A. Emami-Naeini, *Feedback Control of Dynamic Systems*, Reading: Addison-Wesley, 1986.

[Fra78] P.M. Frank, *Introduction to System Sensitivity Theory*, Academic Press, NY, 1978.

[FrD85] B.A. Francis and J.C. Doyle, "Linear control theory with an H^∞ optimality criterion," Systems Control Group Report #8501, Dept. of E.E., Univ. of Toronto, Canada, Oct. 1985.

[Fre85] J.S. Freudenberg, "Issues in frequency domain feedback control," Ph.D. dissertation, Univ. of Illinois, 1985.

[FrL83b] J.S. Freudenberg and D P. Looze, "The multivariable nature of multivariable feedback systems," *Proc. 22nd IEEE Conf. on Decision and Contr.*, San Antonio, TX, Dec. 1983.

[FrL85a] J.S. Freudenberg and D.P. Looze, "Right half plane poles and zeros and design tradeoffs in feedback systems," *IEEE Trans. Auto. Contr.*, vol. AC-30, no. 6, June 1985.

[FrL85d] J.S. Freudenberg and D.P. Looze, "Relations between properties of multivariable feedback systems at different loop-breaking points: part I," *Proc. 24th IEEE Conf. on Decision and Contr.*, Ft. Lauderdale, FL, Dec. 1985.

[FrL86a] J.S. Freudenberg and D.P. Looze, "Relations between properties of multivariable feedback systems at different loop-breaking points: part II," *Proc. 1986 American Contr. Conf.*, Seattle, WA, June 1986.

[FrL86b] J.S. Freudenberg and D.P. Looze, "The relation between open loop and closed loop properties of multivariable feedback systems," *IEEE Trans. on Auto. Contr.*, vol. AC-31, no. 4, April 1986.

[FrL86c] J.S. Freudenberg and D.P. Looze, "An analysis of H^∞-optimization design methods," *IEEE Trans. on Auto. Contr.*, vol. AC-31, no. 3, March 1986.

[FrZ84] B.A. Francis and G. Zames, "On optimal sensitivity theory for SISO feedback systems," *IEEE Trans. on Auto. Contr.*, vol. AC-29, no. 1, Jan. 1984.

[GoV83] G.H. Golub and C.F. Van Loan, *Matrix Computations*, Baltimore, MD, Johns Hopkins Univ. Press, 1983.

[GrH78] P. Griffiths and J. Harris, *Principles of Algebraic Geometry*, NY: Wiley, 1978.

[GrH81] M.J. Greenberg and J.R. Harper, *Algebraic Topology: A First Course*, Reading, MA: Benjamin/Cummings, 1981.

[Gui49] A.E. Guillemin, *The Mathematics of Circuit Analysis*, NY: Wiley, 1949.

[Hal58] P.R. Halmos, *Finite Dimensional Vector Spaces*, Princeton, NJ: Van Nostrand, 1958.

[Hel85] J.W. Helton, "Worst case analysis in the frequency domain: the H^∞ approach to control," *IEEE Trans. Auto. Contr.*, vol. AC-30, 1985.

[HoL84] I. Horowitz and Y.-K. Liao, "Limitations of non-minimum phase feedback systems," *Int. J. Contr.*, vol. 40, no. 5, pp. 1003-1013, 1984.

[Hor63] I.M. Horowitz, *Synthesis of Feedback Systems*, NY: Academic, 1963.

[Hu59] S.T. Hu, *Homotopy Theory*, NY: Academic Press, 1959.

[HuM82] Y.S. Hung and A.G.J. MacFarlane, *Multivariable Feedback: A Quasi-Classical Approach*, Lecture Notes in Control and Information Sciences, vol. 40, Springer-Verlag, 1982.

[Kat82] T. Kato, *A Short Introduction to Perturbation Theory for Linear Operators*, NY: Springer-Verlag, 1982.

[KhT85] P.P. Khargonekar and A. Tannenbaum, "Non-euclidean metrics and the robust stabilization of systems with parameter uncertainty," *IEEE Trans. Auto. Contr.*, vol. AC-30, pp. 1005-1013, Oct. 1985.

[KlL80] V.C. Klema and A.J. Laub, "The singular value decomposition: its computation and some applications," *IEEE Trans. Auto. Contr.*, vol. AC-25, no. 2, pp. 164-176, April 1980.

[Kno56] K. Knopp, *Infinite Sequences and Series*, NY: Dover, 1956.

[Kwa83] H. Kwakernaak, "Robustness optimization of linear feedback systems," *Proc. 22nd IEEE Conf. on Decision and Contr.*, pp. 618-624, Dec. 1983.

[Kwa85] H. Kwakernaak, "Minimax frequency domain performance and robustness optimization of linear feedback systems," *IEEE Trans. on Auto. Contr.*, vol. AC-30, no. 10, Oct. 1985.

[KwS72] H. Kwakernaak and R. Sivan, *Linear Optimal Control Systems*, NY: Wiley-Interscience, 1972.

[LCL81] N.A. Lehtomaki, D.A. Castanon, B.C. Levy, G. Stein, N.R. Sandell, Jr., and M. Athans, "Robustness tests utilizing the structure of modelling error," *Proc. 20th IEEE Conf. on Decision and Contr.*, pp. 1173-1190, Dec. 1981.

[LCL84] N.A. Lehtomaki, D.A. Castanon, B.C. Levy, G. Stein, N.R. Sandell, Jr., and M. Athans, "Robustness and modelling error characterization," *IEEE Trans on Auto Contr.*, vol. AC-29, no. 3, pp. 212-220, March 1984.

[Lee32] Y.W. Lee, "Synthesis of electric networks by means of the Fourier transforms of Laguerre's functions," *J. Math. and Phys.*, vol. XI, pp. 83-113, June 1932.

[LeR70] N. Levinson and R.M. Redheffer, *Complex Variables*, San Francisco: Holden-Day, 1970.

[LSA81] N.A. Lehtomaki, N.R. Sandell, Jr., and M. Athans, "Robustness results in linear quadratic gaussian based multivariable control designs," *IEEE Trans. on Auto. Contr.*, vol. AC-26, no. 1, Feb. 1981.

[Mac79] A.G.J. MacFarlane, ed., *Frequency-Response Methods in Control Systems*, NY: IEEE Press, 1979.

[MaH81] A.G.J. MacFarlane and Y.S. Hung, "Use of parameter groups in the analysis and design of multivariable feedback systems," *Proc. 20th IEEE Conf. on Decision and Contr.*, pp. 1492-1494, Dec. 1981.

[MaH83] A.G.J. MacFarlane and Y.S. Hung, "Analytic properties of the singular values of a rational matrix," *Int. J. Contr.*, vol. 37, no. 2, 1983.

[MaS79] A.G.J. MacFarlane and D.F.A. Scott-Jones, "Vector gain," *Int. J. Contr.*, vol. 29, pp. 65-91, 1979.

[Mun75] J.R.Munkres, *Topology: A First Course*, Englewood Cliffs, NJ: Prentice-Hall, 1975.

[Nyq32] H. Nyquist, "Regeneration theory," *Bell Syst. Tech. J.*, vol. 11, pp. 126-147, 1932.

[OYF85] S.D. O'Young and B.A. Francis, "Sensitivity trade-offs for multivariable plants," *IEEE Trans. on Auto. Contr.*, vol. AC-30, no. 7, pp. 625-632, July 1985.

[OYF86] S.D. O'Young and B.A. Francis, "Optimal performance and robust stabilization," *Automatica*, vol. 22, no. 2, pp. 171-183, 1986.

[PEM81] I. Postlethwaite, J.M. Edwards, and A.G.J. MacFarlane, "Principle gains and principal phases in the analysis of linear multivariable feedback systems," *IEEE Trans. on Auto. Contr.*, vol. AC-26, Feb. 1981.

[PoM79] I. Postlethwaite and A.G.J. MacFarlane, *A Complex Variable Approach to the Analysis of Linear Multivariable Feedback Systems*, Berlin: Springer, 1979.

[Ros74] H.H. Rosenbrock, *Computer-Aided Control System Design,* London: Academic Press, 1974.

[SLH81] M.G. Safonov, A.J. Laub and G.L. Hartmann, "Feedback properties of multivariable systems: the role and use of the return difference matrix," *IEEE Trans. on Auto. Contr.*, vol. AC-26, 1981.

[Spi65] M. Spivak, *Calculus on Manifolds,* NY: Benjamin, 1965.

[Spi70] M. Spivak, *A Comprehensive Introduction to Differential Geometry,* Berkeley: Publish or Perish, 1970.

[Spi79] M. Spivak, *A Comprehensive Introduction to Differential Geometry, Volume II,* Wilmington, Delaware: Publish or Perish, 1979.

[StA84] G. Stein and M. Athans, "The LQG/LTR procedure for multivariable feedback control design," *MIT Lab. for Inform. and Decision Systems,* no. LIDS-P-1384, Cambridge, MA, May 1984.

[Ste73a] G.W. Stewart, *Introduction to Matrix Computations,* NY: Academic Press, 1973.

[Ste84] G. Stein, Notes, ONR/Honeywell Workshop, Oct. 1984.

[Ste85] G. Stein, "Beyond singular values and loop shapes," Report No. LIDS-P-1504, MIT., Cambridge, MA, Aug. 1985.

[VeJ84] M. Verma and E. Jonckheere, "L^∞-compensation with mixed sensitivity as a broad-band matching problem," *Systems and Contr. Letters,* vol. 4, pp. 125-129, May 1984.

[Vic73] J.W. Vick, *Homology Theory,* NY: Academic Press, 1973.

[Vid85] M. Vidyasagar, *Control System Synthesis: A Factorization Approach,* Cambridge, MA: MIT Press, 1985.

[WDH80] J.E. Wall, Jr., J.C. Doyle, and C.A. Harvey, "Tradeoffs in the design of multivariable feedback systems," in *Proc. 18th Ann. Allerton Conf.*, pp. 715-725, Oct. 1980.

[ZaD63] L.A. Zadeh and C.A. Desoer, *Linear System Theory,* NY: McGraw-Hill, 1963.

[ZaF83] G. Zames and B.A. Francis, "Feedback, minimax sensitivity, and optimal robustness," *IEEE Trans. Auto. Contr.*, vol. AC-28, no. 5, May 1983.

[Zam81] G. Zames, ''Feedback and optimal sensitivity: model reference transformations, multiplicative seminorms, and approximate inverses,'' *IEEE Trans. Auto. Contr.*, vol. AC-26, April 1981.

Lecture Notes in Control and Information Sciences

Edited by M. Thoma and A. Wyner

Lecture Notes in Control and Information Sciences

Edited by M. Thoma and A. Wyner

Vol. 81: Stochastic Optimization
Proceedings of the International Conference, Kiew, 1984
Edited by I. Arkin, A. Shiraev, R. Wets
X, 754 pages, 1986.

Vol. 82: Analysis and Algorithms of Optimization Problems
Edited by K. Malanowski, K. Mizukami
VIII, 240 pages, 1986.

Vol. 83: Analysis and Optimization of Systems
Proceedings of the Seventh International Conference of Analysis and Optimization of Systems
Antiba, June 26-27, 1986
Edited by A. Bensoussan, J. L. Lions
XVI, 901 pages, 1986.

Vol. 84: System Modelling and Optimization
Proceedings of the 12th IFIP Conference
Budapest, Hungary, September 2–6, 1985
Edited by A. Prékopa, J. Szelezsán, B. Strazicky
XII, 1046 pages, 1986.

Vol. 85: Stochastic Processes in Underwater Acoustics
Edited by Charles R. Baker
V, 205 pages, 1986.

Vol. 86: Time Series and Linear Systems
Edited by Sergio Bittanti
XVII, 243 pages, 1986.

Vol. 87: Recent Advances in System Modelling and Optimization
Proceedings of the IFIP-WG 7/1 Working Conference
Santiago, Chile, August 27-31, 1984
Edited by L. Contesse, R. Correa, A. Weintraub
IV, 199 pages, 1987.

Vol. 88: Bruce A. Francis
A Course in H_∞ Control Theory
XI, 156 pages, 1987.

Vol. 88: Bruce A. Francis
A Course in H_∞ Control Theory
X, 150 pages, 1987.
Corrected - 1st printing 1987

Vol. 89: G. K. H. Pang/A. G. J. McFarlane
An Expert System Approach to Computer-Aided Design of Multivariable Systems
XII, 223 pages, 1987.

Vol. 90: Singular Perturbations and Asymptotic Analysis in Control Systems
Edited by P. Kokotovic, A. Bensoussan, G. Blankenship
VI, 419 pages, 1987.

Vol. 91 Stochastic Modelling and Filtering
Proceedings of the IFIP-WG 7/1 Working Conference
Rome, Italy, Decembre 10-14, 1984
Edited by A. Germani
IV, 209 pages, 1987.

Vol. 92: L. T. Grujić, A. A. Martynyuk, M. Ribbens-Pavella
Large-Scale Systems Stability Under Structural and Singular Perturbations
XV, 366 pages, 1987.

Vol. 93: K. Malanowski
Stability of Solutions to Convex Problems of Optimization
IX, 137 pages, 1987.

Vol. 94: H. Krishna
Computational Complexity of Bilinear Forms
Algebraic Coding Theory and Applications to Digital Communication Systems
XVIII, 166 pages, 1987.

Vol. 95: Optimal Control
Proceedings of the Conference on Optimal Control and Variational Calculus
Oberwolfach, West-Germany, June 15-21, 1986
Edited by R. Bulirsch, A. Miele, J. Stoer and K. H. Well
XII, 321 pages, 1987.

Vol. 96: H. J. Engelbert/W. Schmidt
Stochastic Differential Systems
Proceedings of the IFIP-WG 7/1 Working Conference
Eisenach, GDR, April 6-13, 1986
XII, 381 pages, 1987.

Lecture Notes in Control and Information Sciences

Edited by M. Thoma and A. Wyner

Vol. 97: I. Lasiecka/R. Triggiani (Eds.)
Control Problems for Systems
Described by Partial Differential Equations
and Applications
Proceedings of the IFIP-WG 7.2
Working Conference
Gainesville, Florida, February 3-6, 1986
VIII, 400 pages, 1987.

Vol. 98: A. Aloneftis
Stochastic Adaptive Control
Results and Simulation
XII, 120 pages, 1987.

Vol. 99: S. P. Bhattacharyya
Robust Stabilization Against
Structured Perturbations
IX, 172 pages, 1987.

Vol. 100: J. P. Zolésio (Editor)
Boundary Control and Boundary Variations
Proceedings of the IFIP WG 7.2 Conference
Nice, France, June 10-13, 1987
IV, 398 pages, 1988.

Vol. 101: P. E. Crouch,
A. J. van der Schaft
Variational and Hamiltonian
Control Systems
IV, 121 pages, 1987.

Vol. 102: F. Kappel, K. Kunisch,
W. Schappacher (Eds.)
Distributed Parameter Systems
Proceedings of the 3rd International Conference
Vorau, Styria, July 6–12, 1986
VII, 343 pages, 1987.

Vol. 103: P. Varaiya, A. B. Kurzhanski (Eds.)
Discrete Event Systems:
Models and Applications
IIASA Conference
Sopron, Hungary, August 3-7, 1987
IX, 282 pages, 1988.

Vol. 104: J. S. Freudenberg/D. P. Looze
Frequency Domain Properties of Scalar
and Multivariable Feedback Systems
VIII, 281 pages, 1988.

GPSR Compliance

The European Union's (EU) General Product Safety Regulation (GPSR) is a set of rules that requires consumer products to be safe and our obligations to ensure this.

If you have any concerns about our products, you can contact us on ProductSafety@springernature.com

In case Publisher is established outside the EU, the EU authorized representative is:

Springer Nature Customer Service Center GmbH
Europaplatz 3
69115 Heidelberg, Germany

Batch number: 10704763

Printed by Printforce, the Netherlands